Book 4

Reproductive Patterns

REPRODUCTION IN MAMMALS

Book 4

Reproductive Patterns

EDITED BY

C. R. AUSTIN
Fellow of Fitzwilliam College,
Charles Darwin Professor of Animal Embryology,
University of Cambridge

AND

R. V. SHORT
Fellow of Magdalene College,
Reader in Reproductive Biology,
University of Cambridge

ILLUSTRATIONS BY JOHN R. FULLER

CAMBRIDGE
AT THE UNIVERSITY PRESS 1972

Published by the Syndics of the Cambridge University Press
Bentley House, 200 Euston Road, London NW1 2DB
American Branch: 32 East 57th Street, New York, N.Y.10022

© Cambridge University Press 1972

Library of Congress Catalogue Card Number: 78–189597

ISBNS: 0 521 08578 0 hardcovers
       0 521 09616 2 paperback

Printed by offset in Great Britain by
Alden & Mowbray Ltd
at the Alden Press, Oxford

Contents

Contributors to Book 4

C. E. Adams
Animal Research Station
307 Huntingdon Road
Cambridge

R. G. Edwards
Physiological Laboratory
Downing Street
Cambridge

J. Herbert
Department of Anatomy
Downing Street
Cambridge

R. M. F. S. Sadleir
Department of Biological Sciences
Simon Fraser University
British Columbia Canada

R. V. Short
Department of Veterinary Clinical Studies
Madingley Road
Cambridge

Preface

Reproduction in Mammals is intended to meet the needs of undergraduates reading Zoology, Biology, Physiology, Medicine, Veterinary Science and Agriculture, and as a source of information for advanced students and research workers. It is published as a series of five small textbooks dealing with all major aspects of mammalian reproduction. Each of the component books is designed to cover independently fairly distinct subdivisions of the subject, so that readers can select texts relevant to their particular interests and needs, if reluctant to purchase the whole work. The contents lists of all the books are set out on the next page.

The contributions in the fourth book describe a number of different *Reproductive Patterns*. A study of unusual species such as the kangaroo and the elephant emphasizes the degree of variation that exists within the general framework of mammalian reproduction. Different patterns of reproductive behaviour are analysed, and some indication is given of the way in which the environment can act to control reproductive activity. Turning to the microenvironment, the ways in which fertility and development can be modified by the immune responses of the body are considered, and the book concludes with a discussion of the problems of reproductive ageing and senescence.

Dr Sadleir wishes to thank A. L. Turnbull for his comments on Chapter 3, and the audio-visual centre of Simon Fraser University for their preparation of the Figures.

Books in this series

1 Species differences
R. V. Short

Our knowledge of reproductive physiology has been built up from a detailed study of a mere handful of the world's 5000 species of mammals, and yet we still naively believe that all mammals must conform to a common plan. Species differences tend to baffle us, because they provide exceptions to our man-made rules, and we are apt to think of them as quirks of nature, best excluded from our thoughts lest they confuse us. But we are clearly in no position to make generalizations and draw up rules which all mammals are likely to obey.

The general pattern of *Reproduction in Mammals* reflects the fact that it is customary to teach reproductive physiology on a systems basis, with minor asides to acknowledge the more notable species differences. But in order to provide a warp for this weft, we occasionally need to adopt a species approach, to see how all these complex reproductive mechanisms are woven into the life cycles of individual animals. So much has already been written about the reproduction of man and labora-tory animals that it would be rather tedious to go over the same ground again, merely from a different viewpoint. In this chapter we will therefore consider some of the more unexpected reproductive mechanisms in some of the more unusual species.

THE MARSUPIALS

There is a wide variety of marsupials, indigenous to North and South America (70 species) and Australia (160 species) (see Fig. 1-1). But man has lost no time in spreading them around the world, so that, for example, wallabies have even become feral in certain parts of England, and the brush possum is a veritable pest in New Zealand, where it does extensive damage to the

I

forests. We owe much of our recent knowledge about marsupial reproduction to the outstanding work of Geoff Sharman and his colleagues in Australia, and it is indeed a fascinating story.

All marsupials have in common the fact that the young are born in a very immature state of development after an extremely short gestation period. Nevertheless, at birth they have to make their way unaided into the pouch, or marsupium, where they

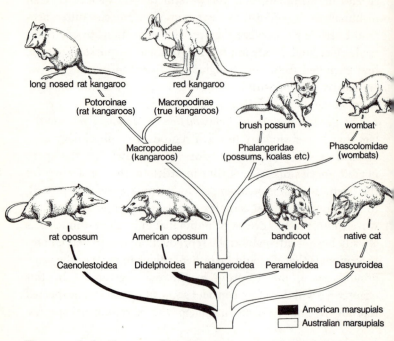

Fig. 1-1. A family tree of present-day marsupials in America and Australia. (From G. B. Sharman. *Science J.*, March 1967, p. 55.)

will continue their development whilst permanently attached to a teat (see Fig. 1-2). To deliver her young, the red kangaroo, like almost all other marsupials, sits down with her tail between her legs, and licks the fetal fluids and placenta as they trail behind the slowly crawling fetus. She definitely does not lick a path in front of it to guide it into the pouch, as people had previously supposed.

The red kangaroo is the largest marsupial, and an adult male may stand 6 ft high and weigh as much as a man. It also produces the largest young at birth, but even so the 'joey' only weighs 0.8 g. The smallest newborn marsupial, belonging to one of the native cats, weighs a mere 12 *mg*, and it is scarcely credible that so small an animal could be capable of independent locomotion. These pouch young are little more than exteriorized

Fig. 1-2. A young red kangaroo joey in the pouch, firmly attached to one of the teats. The large teat on the left is still being suckled by the young kangaroo that has recently left the pouch. (From G. B. Sharman and J. H. Calaby. *C.S.I.R.O. Wildlife Research* **9**, 58, plate 5, fig. 2 (1964).)

fetuses, and one wonders why physiologists have not made more use of them for the study of fetal physiology.

The female reproductive tract of marsupials is quite unlike that of eutherian ('higher') mammals: there are two lateral vaginae, and a central pseudovaginal canal, which is usually closed until the approach of parturition (see Fig. 1-3). At mating, spermatozoa pass up the lateral vaginal canals to enter the two

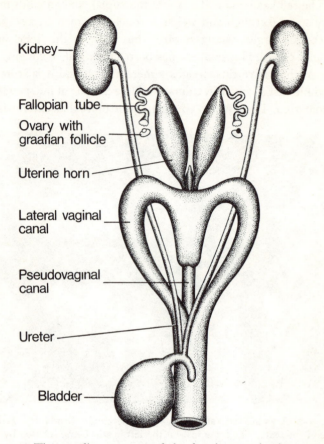

Kidney

Fallopian tube

Ovary with
graafian follicle

Uterine horn

Lateral vaginal
canal

Pseudovaginal
canal

Ureter

Bladder

Fig. 1-3. The peculiar anatomy of the female reproductive tract in marsupials. Spermatozoa ascend by way of the lateral vaginal canals, and the fetus is born through the central pseudovaginal canal, which is occluded at other times. (From G. B. Sharman. *Science J.*, March 1967, p. 54.)

separate uterine horns, and partition of the ejaculate between these two alternative pathways is aided in some species by the fact that the male has a bifurcated penis. Reproductive duplication may not end there, because in many marsupials the spermatozoa swim around in pairs, joined together by their

Fig. 1-4. The conjugated spermatozoa of the Virginia opossum, showing the way they are attached to one another by the acrosome.

acrosomes (see Fig. 1-4). This pairing process takes place in the epididymis, but its function is unknown.

At the time of parturition, the blocked central pseudovaginal canal must open to permit the fetus to pass through; birth does not seem to occur by way of the lateral vaginal canals. This process is reminiscent of the relaxation of the cervix and pelvic ligaments in eutherian mammals, which is thought to be under the control of the hormone relaxin. However, limited evidence in marsupials suggests that recanalization is controlled by progesterone from the corpus luteum.

Since marsupials have such short intrauterine gestations, it is not surprising that the endocrinology of pregnancy is much simpler than in eutherian mammals. The placenta in most marsupials is of the primitive yolk-sac type, and in no marsupial does it acquire any endocrine activity; the mere presence of a fetus in the uterus therefore cannot upset the length of the normal oestrous cycle. Marsupials have never been faced with the need to develop a mechanism for the endocrine recognition of pregnancy, nor have they developed the eutherian trick of terminating the life of the cyclical corpus luteum by luteolytic activity of the uterus. It is curious that the time of parturition seems to bear no consistent relationship to the length of the oestrous cycle; one would have imagined that the most convenient way of expelling the fetus from the uterus would have been to rely on progesterone withdrawal and the oestrogen-induced uterine contractility that occurs when the animal returns to oestrus. But, depending on the species, the gestation period may be longer than, the same length as, or even shorter than the oestrous cycle (see Fig. 1-5). The logical conclusion is that neither oestrogen nor progesterone has anything to do with parturition, so one wonders what does cause it, and how the mother becomes aware that her minute fetus is about to enter the world.

Although an embryo in the uterus has no effect on the length of the oestrous cycle, the presence of a joey in the pouch, attached to one of the teats, has a major effect. The inhibition of ovarian activity by lactation is of course well recognized in many eutherian mammals from mice to man, but in marsupials the suckling stimulus has been developed as the sole means by which the developing youngster can modify the reproductive cycles of its mother. Some such feedback control is essential, because, in the red kangaroo for example, the joey needs to spend about 235 days attached to a teat before it can complete its pouch development and be free to come and go as it chooses. Since the length of intrauterine gestation is only 33 days, it would be a positive embarrassment if a new youngster arrived in the pouch every month, only to find no teat accommodation available.

In most marsupials, the length of the gestation period is fortunately much shorter than the oestrous cycle. In these animals, the suckling stimulus of the newborn pouch young

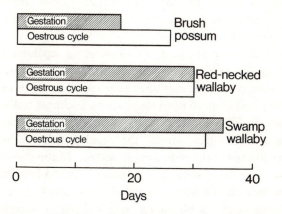

Fig. 1-5. The relationship between gestation length and oestrous cycle length in three marsupials. In the brush possum, post-partum oestrus is inhibited by the suckling stimulus of the pouch young. In the wallabies, the young do not reach the pouch in time to prevent oestrus, so they have evolved a lactation-induced embryonic diapause.

therefore has time to exert a direct inhibitory effect on the hypothalamus, so that pituitary gonadotrophin release is impaired. Thus the animal fails to return to oestrus at the expected time, and will remain in anoestrus with inactive ovaries whilst the joey is on the teat. If the joey dies, or is removed from the pouch, the animal returns to oestrus within a few days.

Problems begin to arise in the macropod marsupials (see Fig. 1-1), such as the red kangaroo, quokka, potoroo, and the red-necked, tammar and swamp wallabies, where gestation length is similar to, or longer than, the length of the oestrous cycle (see Fig. 1-5). In this group of animals, the joey does not reach the teat soon enough to prevent a return to oestrus, so the animal mates and conceives again. The peculiar anatomy of the female reproductive tract comes into its own here. The one-way traffic for ascending sperm via the lateral vaginal canals, and for the descending fetus in the central vaginal canal, together with the fact that ovulation alternates between the two ovaries, means that a new conception can occur when the animal is still pregnant with a fetus in one uterine horn. Indeed, parturition and mating could occur almost simultaneously if the need arose.

Faced with the prospect of two offspring separated in gestational age by only about 30 days, these macropod marsupials have evolved an ingenious device for arresting the development of the second embryo until the first one is almost ready to leave the pouch. The suckling stimulus of the young joey, newly attached to the teat, arrests the development of the corpus luteum formed at the post-partum oestrus. Mating and fertilization occur at this oestrus, but in the absence of a fully functional corpus luteum the uterus is unable to support the continued development of the blastocyst, which therefore enters a phase of arrested development or 'embryonic diapause'. This occurs soon after the blastocyst has entered the uterus, when it is just a hollow sphere of about 100 undifferentiated cells surrounded by a shell membrane; all mitoses stop, and in the red kangaroo the blastocyst remains in this dormant state for just over 200 days before a normal rate of growth is suddenly resumed.

The role of the ovaries in the maintenance of diapause is entirely passive, since the blastocyst is unaffected by ovariectomy. But the ovaries are necessary for reactivation of the blastocyst, and this effect can be mimicked by injections of progesterone, or occasionally by injections of oestrogen. It seems likely that the suckling stimulus of the pouch young acts by producing a reflex release of oxytocin from the maternal posterior pituitary, which in turn has a direct inhibitory action on the corpus luteum (see Fig. 1-6). If the joey dies, or is

8

removed from the pouch, normal embryonic development is resumed almost immediately so that a new individual is born about 31 days later.

Throughout this period of embryonic diapause, when there is an inactive corpus luteum in one ovary, the kangaroo also fails to return to oestrus. One would have imagined that this

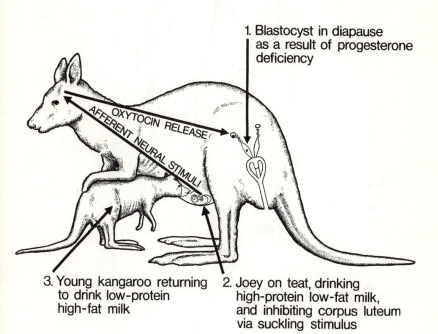

1. Blastocyst in diapause as a result of progesterone deficiency

3. Young kangaroo returning to drink low-protein high-fat milk

2. Joey on teat, drinking high-protein low-fat milk, and inhibiting corpus luteum via suckling stimulus

Fig. 1-6. At any one time, a female red kangaroo may have three young of different ages dependent on her.

was due to a direct inhibitory effect of the suckling stimulus on gonadotrophin release. But curiously, the explanation is more complex than that; if the so-called 'inactive' corpus luteum of diapause is surgically removed from the ovary of an animal with a joey in the pouch, oestrus ensues. Therefore we must conclude that the corpus luteum and not the suckling stimulus normally prevents these animals coming into oestrus during embryonic

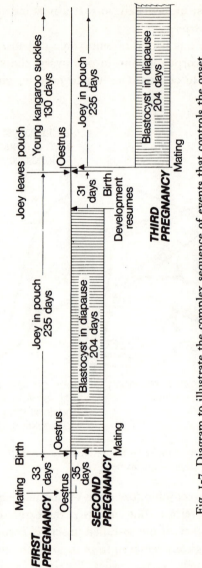

Fig. 1-7. Diagram to illustrate the complex sequence of events that controls the onset and termination of embryonic diapause in the red kangaroo.

diapause. The corpus luteum is therefore not as inactive as we had supposed.

An attempt has been made to summarize this complicated state of affairs in Fig. 1-7. It can be seen that a female red kangaroo may have three youngsters dependent on her at any one time, all in different stages of development. The first oestrus of the breeding season will be followed by a normal intrauterine gestation period of 33 days, resulting in the arrival of the first joey in the pouch. Oestrus occurs 2 days later, and it is the presence of the joey on the teat that causes the second embryo to have an enormously prolonged intrauterine gestation of about 235 days, most of which is occupied by the period of embryonic diapause. The intensity of the suckling stimulus provided by the first youngster begins to wane towards the end of its pouch life, when it is making tentative forays into the outside world. This allows the corpus luteum to become re-activated, so that the delayed blastocyst in the uterus can resume its development. The whole process is perfectly timed, so that about a day after the first youngster forsakes the pouch permanently, a second one arrives to take its place. The first youngster is not finally weaned for a further 4 months, and continues to put its head into the pouch at meal-times, drinking from the same enlarged teat to which it was formerly attached. By now the milk from this mammary gland has changed its composition to become rich in fat and poor in protein. Meanwhile, the new joey has attached itself to one of the other teats, which provides it with a high-protein, low-fat milk – a remarkable example of a division of function between the various parts of the mammary gland.

Within a day of the birth of the second youngster and its entry into the pouch, the kangaroo returns to oestrus again and is mated. As you might expect, the development of the third embryo is also arrested by the suckling efforts of the second one in the pouch, so like its predecessor it will experience a prolonged intrauterine gestation of about 235 days.

Many eutherian mammals such as bats, bears, seals and badgers also have a phase of arrested blastocyst development (see Book 1, Chapter 4), with the object of ensuring that the young are born at the most favourable time of the year. But in the macropod marsupials, embryonic diapause seems to have a rather different function. Young can be safely born at any time of the year, because they will be protected from the rigours of the external environment by the pouch. But the irregular periods of drought, typical of certain parts of Australia, may so debilitate the kangaroo that she is unable to provide the ever-increasing quantities of milk demanded by the growing pouch young. As a result, the joey may die; in this event, the embryo in diapause immediately resumes development, so that the mother soon receives a small replacement joey in her pouch that will not initially be so demanding on her depleted reserves. Embryonic diapause therefore serves as a sort of reproductive 'spare tyre'.

Marsupial reproduction has provided us with so many surprises in the past that there must be more in store for us as we begin to investigate more species. Some of their reproductive mechanisms are so unusual as to suggest that viviparity has evolved completely independently in eutherian and marsupial mammals. The fact that both groups control their reproductive processes by the same steroidal sex hormones suggests that these compounds indeed have a long ancestry.

ELEPHANTS

'There is an animal called an elephant . . . no larger animals can be found. They possess vast intelligence and memory. And they copulate back to back. Elephants remain pregnant for two years, nor do they have babies more than once, nor do they have several at a time, but only one. They live three hundred years.' This mixture of fact and fiction is gleaned from a twelfth-century Latin bestiary; the ancients made a particular study of both Asiatic and African elephants because of their importance as beasts of burden, and as tactical weapons in warfare. It is incredible that even in 350 B.C., Aristotle was well aware of the permanent intra-abdominal location of the testes, and he correctly recorded the gestation length.

There has recently been a resurgence of interest in elephants, particularly the larger African elephant, because of the enormous amount of damage that these animals can do within the confines of a National Park. With increasing human population pressure and land hunger throughout Africa, the natural range of elephants has been severely restricted, so that large numbers of animals have now become crowded into areas that are too small to support them. This has created an apparent elephant population explosion, and the animals have started to destroy their habitat, turning thick forest into open savannah. This has unfortunate consequences for many smaller mammals and birds that can only live in a forest environment; even the elephants themselves may be adversely affected, as they are predominately browsing animals and may require some protection from the heat of the tropical sun. In an attempt to reduce the elephant population to a figure more in line with the carrying capacity of the habitat, many National Parks have been forced to kill them in large numbers. Such game cropping operations have provided scientists with a wealth of fascinating biological material, and the work of Dick Laws in Kenya, Uganda and Tanzania, and John Hanks in Zambia, has added much to our knowledge of the species, and has allowed us to see whether elephants are able to regulate their reproductive rate and so adapt to increasing population pressure.

Table 1-1 suggests that both the age at puberty and the interval between the birth of successive calves are indeed density-dependent variables, so that elephants might ultimately be able to avoid the problem of overpopulation. But in such a long-lived species, many years would elapse before these natural checks on the birth rate had a significant effect on numbers, by which time the habitat might have been irrevocably damaged.

Before we can begin to estimate the annual rate of growth of an elephant population, we must calculate mortality rates for

Species differences

the different age classes, and know the total life span. Calculating accurate mortality rates for such a long-lived species is almost impossible, since it involves making the assumption that in the past the population has remained in a stable equilibrium state, with no sudden changes. The only significant predator of the elephant is man, and his changing role in the last 50

TABLE 1-1. Evidence for density-dependent controls on the reproductive rate of female African elephants

Locality	Population density (elephant/sq. mile)	Age at puberty (years)	Calving interval (years)
Tanzania (Mkomazi)	2–3	12	3
Zambia (Luangwa)	3–6	14	4
Kenya (Tsavo)	3	15	7
Uganda (N. Murchison)	3–4	16	9
Uganda (S. Murchison)	6–7	18	6*
Uganda (Budongo Forest)	6–10	22	Not known

* If there is a high neonatal mortality, the calving interval may decline because of the shortened period of lactational anoestrus.

years from hunter to conservationist must itself have disturbed this equilibrium.

If these limitations are accepted, we can construct a 'survivorship curve' by collecting and ageing all the jaws of elephants that have died in the area from natural mortality. One can get a good idea of an elephant's age from the stage of eruption and state of wear of its six pairs of molar teeth in the lower jaw; by the age of 60, the last of these teeth is worn down to the gum, thus placing a finite limit on the animal's life span.

Whilst not even a hyaena can destroy the lower jaw of an adult elephant until it has been weathered by sun and rain for several years, little remains of the skull of a young elephant even within a few months of its death. So survivorship curves based on found jaws will seriously under-represent the mortality of the younger age classes, which are those that are suffering the heaviest losses. But if we know the age at puberty, the calving interval and the age structure of the adult population, we can calculate the number of calves that should be born, and so arrive at an indirect estimate of neonatal mortality.

Survivorship curves can also be constructed by examining the age structure of a truly representative random sample of the existing population, once again based on the dubious assumption that the population is in a stable equilibrium state. The only way to age a random sample of the standing population is to kill at random, taking entire family units of elephants irrespective of age, sex or condition. Such an apparently heartless procedure is naturally difficult to justify to an uninformed lay public, Parks administration or Government.

Fig. 1-8 shows survivorship curves for female elephants in Kenya calculated by Dick Laws from found jaws, and for female elephants in Zambia calculated by John Hanks from a sample of the population cropped at random. These curves follow a cohort of 1000 animals through from birth to death, and they emphasize the magnitude of the mortality during the first 10 years of life, which could itself be density-dependent. Undoubtedly this heavy neonatal mortality, more than any change in the age at puberty or the duration of lactational anoestrus, is the main factor regulating population growth, and the same is probably true of the populations of all wild animals; it is certainly true of the human population.

Armed with these survivorship curves, and the data on the age at puberty and calving interval in Table 1-1, some predictions can be made with a computer about the rate of elephant population growth. These predictions must be very tentative because of the assumptions made in the calculation of the survivorship curves, but nevertheless they are better than predictions based on the crude birth rate, which are utterly meaningless although still frequently used.

The computer suggests that the Kenyan elephant population is just about stationary, and that the Zambian population is

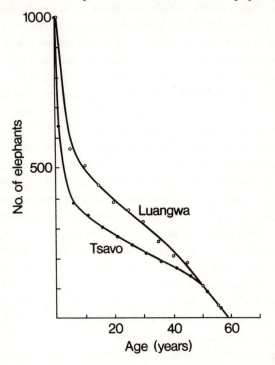

Fig. 1-8. Survivorship curves for female elephants in the Tsavo National Park, Kenya, calculated from found jaws, and in the Luangwa Valley, Zambia, calculated from a cropped sample of the population.

increasing slightly, emphasizing that the 'elephant population explosion' in both these areas is not due to an increased reproductive rate, but rather to overcrowding as a consequence of human encroachment and restriction of the animal's natural range. Cropping may be essential to reduce numbers to within the carrying capacity of the habitat, but if the population is only just holding its own, one must crop with care.

One beneficial result of the cropping operations is that they have allowed us to build up a more complete picture of the reproductive physiology of the elephant. The 22-month gestation period is by far the longest of any mammal, and the maternal ovaries are unusual since they contain large numbers of corpora lutea throughout gestation. Initially, accessory ovulations were thought to occur during pregnancy, but careful counts of corpora lutea at all stages of gestation have shown that this is not the case; in fact, there tend to be more corpora lutea present in the very early stages than later on. The elephant is also peculiar in showing a great variability in the size of its corpora lutea, which may range from 4 mm to 4 cm in diameter. We were astonished to find that the corpora lutea are almost totally devoid of progesterone, and that no progesterone can be detected in the peripheral blood of elephants at any stage of gestation. However, the corpora lutea must possess some sort of endocrine activity, because the degree of development of the endometrial glands in the uterus is directly related to the mass of luteal tissue present in the ovaries.

Several years ago, I was able to follow a wild African elephant through most of its 3-day period of oestrus, and when it was shot at the end of this time, I was surprised to find a number of well-developed corpora lutea in the ovaries, together with a single fresh ovulation point. This suggested that elephants must be monovular, though polyoestrous, and that they must accumulate the corpora lutea from several oestrous cycles until they eventually become pregnant. Unfortunately we do not know the length of the oestrous cycle in the African elephant; that of the Asian elephant is about 3 weeks. But something seemed to be wrong somewhere; there may be as many as 40 corpora lutea present during early pregnancy, but this could not possibly mean that elephants undergo forty 3-week oestrous cycles before conceiving. The relative paucity of non-pregnant elephants with corpora lutea in their ovaries suggests that the elephant must actually spend a relatively short period of time undergoing oestrous cycles before it gets pregnant. If it is to accumulate a large number of corpora lutea in so short a space of time, this must mean that oestrus is frequently accompanied by multiple ovulations (see Fig. 1-9).

The animal seems to need to accumulate a certain critical mass of luteal tissue in its ovaries before the endometrium becomes sufficiently developed to permit a pregnancy. In this respect, elephants are unique in their reproductive inefficiency, and one would love to know why their corpora lutea are so inactive,

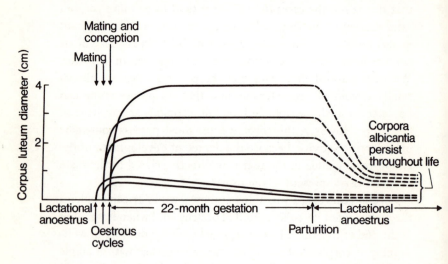

Fig. 1-9. The reproductive pattern of the female elephant, showing how a number of oestrous cycles precede pregnancy, and how corpora lutea are accumulated from cycle to cycle.

why they vary so much in size, and what hormones they really secrete. Since the remnants of the corpora lutea persist in the ovaries throughout life as brown 'corpora albicantia', one can always estimate the number of ovulations that have occurred since puberty. Each pregnancy also leaves behind a permanent scar in the uterus where the placenta was attached, so one can also count the number of calves that an animal has had during its lifetime. Both these techniques can provide valuable information about the reproductive history of the population.

Another curious feature about elephants is that they continue to grow throughout most of their life. Fig. 1-10 shows that the initial growth rates of male and female African elephants are very similar until about the time of puberty, when they diverge. The body weights and shoulder heights of females begin to level off by about the age of 40, whereas males keep on growing at a relatively rapid rate; the testes also continue to enlarge, and the seminiferous tubule diameters keep on increasing with age.

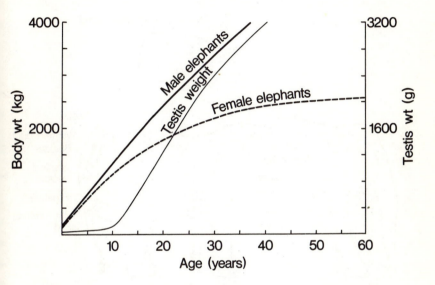

Fig. 1-10. Growth rates of male and female African elephants, showing the marked sexual dimorphism that starts at puberty, when the testes begin to enlarge. In the male, testis weight and body weight increase throughout life.

The reproductive physiology of the male elephant poses a number of interesting problems, and Fig. 1-11 illustrates why there are so many fanciful legends about the manner in which elephants copulate. The vagina of the female opens ventrally, so that the male has to hook the end of his penis into it. Probably only a few inches of the penis actually enter the vagina, so if spermatozoa are to reach the uterus, which is several yards away, it is essential to have a large ejaculate volume. The male accessory organs are therefore extremely well developed, especially the seminal vesicles and the bulbo-urethral glands.

Elephants are one of the few mammals with permanent intra-abdominal testes, and since the mean deep body temperature is about 37 °C, the seminiferous epithelium must be particularly temperature-resistant. Since the elephant cannot keep

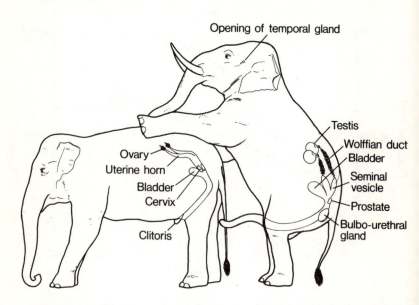

Fig. 1-11. Mating Asiatic elephants, showing the relative positions of the male and female reproductive tracts.

his testes cool, there has been no development of the pampiniform plexus, the countercurrent heat exchange mechanism between spermatic artery and vein found in all scrotal mammals. Elephants are also peculiar in having no discrete epididymis; instead, an extremely tortuous and convoluted duct connects the testes to the openings of the seminal vesicles. In scrotal mammals, spermatozoa are stored in the epididymis, which, like the testis, is kept below body temperature; but the elephant has to store his sperm, probably for long periods of time, at normal body temperature.

Much speculation has surrounded the phenomenon of 'musth', which has been recognized ever since man first started to domesticate male Asiatic elephants, thousands of years ago. It is a time when adult bulls become disobedient, aggressive, and extremely dangerous, often attempting to kill their mahouts, or anybody else who comes within range. Therefore they have to be fettered by stout chains until the mood passes after a month or so. The recent studies of Jainudeen and his colleagues in Ceylon have shown that musth occurs fairly regularly once a year in all healthy adult males, and although it may be brought on prematurely by repeatedly exposing a bull to oestrous females, bulls will mate at any time of the year, irrespective of whether or not they are in musth. Fortunately the mahout gets some warning of impending musth, because the temporal gland starts to swell (see Fig. 1-11). Eventually it begins to discharge a strong-smelling watery secretion down the side of the face, and when in full musth urine dribbles almost continuously from the tip of the penis. We have shown that musth is associated with an enormous increase in the blood testosterone level, which may account for the aggressive behaviour, but there is little information about the behavioural significance of musth in elephants in the wild state. African elephants, which in many ways are so similar to their Asiatic cousins, do not seem to show musth, and their temporal glands secrete copiously irrespective of age, sex or season.

THE HORSE AND ITS HYBRIDS

'I know two things about the horse, and one of them is rather coarse', or so runs the ditty. The horse has in fact played a central role in the development of our knowledge of mammalian reproduction, for it was the phenotypic appearance of the mule (donkey ♂ × horse ♀) that first persuaded scientists and philosophers that *both* parents must contribute to the makeup of the offspring, contrary to the Aristotelean view that this was uniquely the prerogative of the male. The discovery by Harold Cole in 1930 that the blood of pregnant mares contained high levels of gonadotrophin provided us with the first readily available commercial source of follicle stimulating hormone, and PMSG (pregnant mares' serum gonadotrophin) is still widely used for this purpose. When Bernhard Zondek announced in 1934 that stallions excreted oestrogens in their urine, this was the first intimation that sex differences in hormone production were relative rather than absolute, quantitative rather than qualitative. These facts alone should justify some mention of the horse in any textbook of reproductive biology, and a number of recent discoveries have strengthened the case still further.

Perhaps the most remarkable of these was made independently by Dr van Niekerk in South Africa and Professor Bielanski in Poland. They found that only *fertilized* horse eggs ever manage to reach the uterus; unfertilized ones remain trapped at the isthmus of the Fallopian tube, where they often undergo some degree of parthenogenetic cleavage before slowly degenerating during the ensuing months (see Fig. 1-12). If a mare has a succession of sterile oestrous cycles, followed by one at which mating takes place, the developing morula can in some miraculous way jump the queue of unfertilized eggs waiting in the tube, and enter the uterus. How this is achieved remains a complete mystery.

Although it is well recognized that the *fetal* placenta can produce gonadotrophic hormones during pregnancy in many species (see Book 3, Chapters 1 and 3), the horse has always appeared to be an exception to the general rule; PMSG is known to be produced by ulcer-like structures on the inner surface of the uterus, the endometrial cups, and hence it has been assumed that it is a *maternal* hormone. The recent studies of Twink Allen and his colleagues in Cambridge have now shown conclusively that the endometrial cup tissue is fetal in origin after all, and its mode of formation and regression are of great fundamental interest.

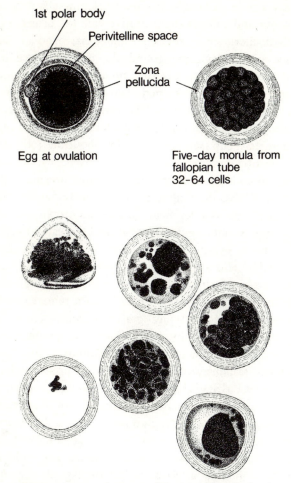

1st polar body

Perivitelline space

Zona pellucida

Egg at ovulation

Five-day morula from fallopian tube 32-64 cells

Degenerating ova from one fallopian tube, ranging in age 24h - $7\frac{1}{2}$ months

Fig. 1-12. A normal horse egg before and after fertilization, and degenerating tubal eggs. (From C. H. van Niekerk and W. H. Gerneke. *Onderstepoort J. Vet. Res.* **33**, 195, fig. 29 (1966).)

Species differences

The early horse conceptus retains its spherical shape on entering the uterus, and by the end of a month it has grown to about the size of an orange. As the developing allantois begins to grow out and convert the primitive yolk-sac placenta into a true allantochorionic placenta, an equatorial band or allanto-chorionic girdle develops around the circumference of the embryo (see Fig. 1-13). Starting on about the 35th day of gestation, cells become detached from this region, to penetrate the maternal endometrium and burrow deep into the stroma, where they enlarge to form the characteristic 'decidual cells' of the endometrial cup (see Fig. 1-14). On about the 40th day, the cup tissue first becomes visible to the naked eye as a band running around the circumference of the pregnant uterine horn, and PMSG first appears in the maternal circulation. Endometrial cup development and gonadotrophin levels increase to a maximum by about day 60, and thereafter the gonadotrophin titres begin to fall as the decidual cells become surrounded by a mass of lymphocytes (see Fig. 1-14). Eventually, the cup tissue is sloughed off from the surface of the uterus into the uterine lumen, a process which bears a remarkable histological resemblance to a typical graft rejection reaction.

If a mare is covered by a jack donkey to produce a mule fetus, the hybrid endometrial cups stimulate an enhanced lymphocytic response, and are sloughed off prematurely, with the result that PMSG is barely detectable in the maternal circulation although the pregnancy is normal in all other respects. Thus it seems that gonadotrophin production is determined by the genotype of the fetus; the mother apparently becomes 'aware' of the fetal cells that have invaded her endometrial stroma and mounts a classical immunological response to reject them. Generally speaking, fetal tissue seems to be completely protected from maternal immunological attack (see Book 2, Chapter 1), so that the endometrial cups of the horse provide an intriguing exception to the rule.

One of the unanswered questions is: why does the horse need to produce all this gonadotrophin in any case? There may well

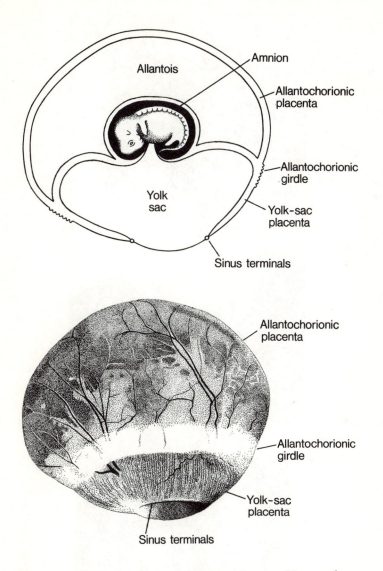

Fig. 1-13. Appearance of a normal horse embryo and its membranes in section and in surface view on the 35th day of gestation. The conceptus is about the size of an orange, and cells become detached from around the allantochorionic girdle at this stage to invade the endometrium and form the endometrial cups.

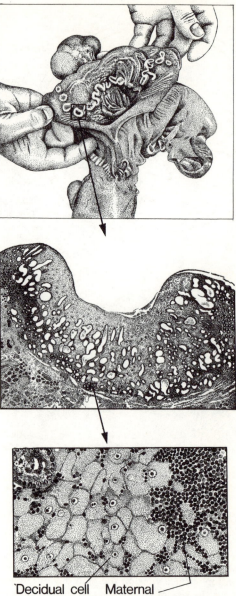

'Decidual cell   Maternal
leucocytic response

be over a million international units of PMSG present in the circulation at any one time, and yet this massive amount only occasionally causes hyperstimulation of the maternal ovaries. But it does result in the formation of a number of accessory corpora lutea, either by ovulation or by luteinization of un-ruptured follicles, and they may act as a supplementary source of progesterone. If a mare aborts after the 35th day of gestation, when the fetal cells have already invaded the endometrium, the cups continue to develop normally and secrete PMSG, and the corpora lutea that are present may be maintained. This suggests that PMSG may have a luteotrophic action, which one might expect of a substance that seems to possess both FSH- and LH-like activities in the one molecule.

By mid pregnancy, all the corpora lutea have regressed and the maternal ovaries no longer produce a significant amount of progesterone. This function is now assumed by the placenta, which contains high concentrations of the hormone; the levels are also high in the fetal circulation, but progesterone virtually disappears from the peripheral blood of the mare herself. This can be explained by the anatomy of the equine placenta, which is of the diffuse epithelio-chorial type. Since it is attached over the entire surface of the endometrium, placentally produced steroids can diffuse directly into their target organ, the uterine tissues, without first having to enter the maternal circulation. They are metabolized by the uterus and elsewhere, and the inactive metabolites are subsequently excreted in the maternal urine.

Fig. 1-14. *Upper* Appearance of the inner surface of a mare's uterus on the 60th day of gestation after removal of the embryo and its membranes. The ring of endometrial cup tissue is clearly visible.

Middle Low-power section through an endometrial cup, showing distended uterine glands filled with secretion rich in PMSG activity. The area between the glands is packed with 'decidual' cells.

Lower High-power section near the base of the cup to show the 'decidual' cells which are fetal in origin, and which produce the PMSG. Their presence has elicited a strong maternal lymphocytic response which will eventually result in the 'rejection' of the endometrial cup.

During the later months of pregnancy, the mare excretes large amounts of oestrogens in her urine, including two unusual compounds, equilin and equilenin (see Fig. 1-15). These oestrogens, which are characterized by having double bonds in ring B, are only found during pregnancy, and are peculiar to the horse family. They are of considerable interest to biochemists, because they are the only known steroids synthesized in the body by a pathway that does not involve cholesterol as an intermediate. The gonads of the horse fetus, whether testes or

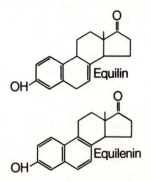

Fig. 1-15. Structural formulae of equilin and equilenin, two ring-B-unsaturated oestrogens with low biological activity that are only found in the urine of the pregnant mare, having been synthesized in the fetus by a cholesterol-independent pathway.

ovaries, undergo a remarkable hypertrophy towards the end of pregnancy, becoming even bigger than the maternal ovaries, and since fetal gonad size and maternal urinary oestrogen excretion parallel one another, it is possible that the two events are in some way related.

Parturition in the mare is also a notable event. In Book 2, Chapter 3, we saw how the activity of the fetal adrenal glands seems to determine the time of birth, suggesting that parturition is under fetal rather than maternal control. In mares, however, parturition almost invariably occurs at night; we timed 501 foalings at thoroughbred studs in Newmarket, and found that 86 per cent occurred between the hours of 7 p.m. and 7 a.m., with a peak incidence around midnight. So whilst the fetus may determine the day of birth, there must be some superimposed maternal fine adjustment mechanism which decides the actual hour.

We have already made some mention of the mule – that unfortunate animal, begotten unnaturally, devoid of an evolutionary history, and denied the ability to reproduce its kind. A number of other equine hybrids have also been produced, including the hinny (horse ♂ × donkey ♀) and a variety of horsebras, zebrorses and zebronkeys (see Fig. 1-16). Although once regarded as mere curiosities, we are now beginning to realize that they are of great scientific interest; horses, donkeys and zebras all have different chromosome numbers, and the hybrids therefore have numbers intermediate between those of their two parents. Whilst this does not present any difficulties to hybrid cells undergoing mitosis, problems begin to arise in the germ cells as they enter meiosis. The first stage of a meiotic division involves the pairing of homologous chromosomes (see Book 1, Chapter 2), and if the maternal and paternal chromosome sets are dissimilar in size and number, true pairing is impossible. Thus the testes of male hybrids secrete normal amounts of testosterone, but show a block in spermatogenesis at about the pachytene stage of meiosis, so that no spermatozoa are formed (see Fig. 1-17). The situation in the ovary of the female hybrid is somewhat different, because the germ cell has to enter meiosis before it is capable of inducing the development of a normal Graafian follicle, and it is the follicle that is the endocrine apparatus of the ovary. The ovaries of hybrids therefore contain very few follicles, and although it is true that some mules can occasionally come into oestrus and actually ovulate, this is a rather rare event.

Whether or not mules are ever fertile has been a lively topic of discussion for centuries. Although there are a number of recorded instances of mules seen with a foal at foot, these cannot really be accepted as evidence. In the first place, the mother

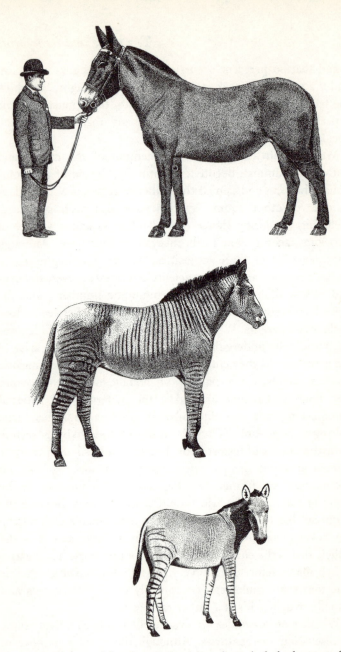

Fig. 1-16. *Upper* 16-hand mule bred by a large jack donkey out of a carthorse mare.
Middle Zebrorse bred by a Grevy's zebra stallion out of a mare.
Lower Zebronkey bred by a common zebra stallion out of a jenny donkey.

Karyotype of male zebronkey
n=53

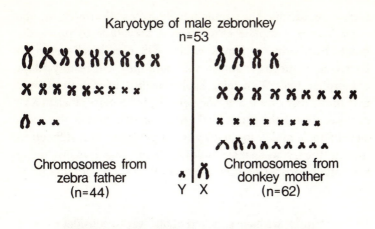

Chromosomes from
zebra father
(n=44)

Y   X

Chromosomes from
donkey mother
(n=62)

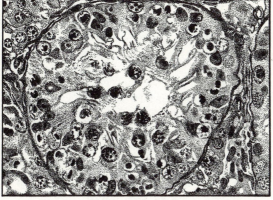

Zebronkey testis

Fig. 1-17. *Upper* Karyotype of a male zebronkey, showing the chromosomes contributed by each of the parents, and illustrating the lack of pairs of homologous chromosomes.
Lower Failure of spermatogenesis in the testis of a male zebronkey. The large cells near the centre of the seminiferous tubule are primary spermatocytes, arrested at the pachytene stage of meiosis as they try in vain to pair homologous chromosomes in preparation for the first reduction division.

cannot be identified as a mule merely by visual inspection; one must have a complete chromosomal karyotype to prove the point. Furthermore, female mules will occasionally start to lactate spontaneously, and in this state they will readily adopt any orphan foal. With remarkable insight, Haldane postulated that in the case of interspecific hybrids, the heterogametic (XY) sex was always the more likely to be sterile. Whilst the occasional female mule can shed an egg at ovulation, it is difficult to imagine the testis of a male mule ever being capable of producing the millions of spermatozoa necessary for a successful insemination.

In this chapter, we have gone out of our way to consider some of the more unusual reproductive mechanisms in a variety of unusual mammals. If the diversity of Nature has impressed you, you will perhaps appreciate how little we can know of man, if only man we know.

SUGGESTED FURTHER READING

The red kangaroo. G. B. Sharman. *Science Journal* (March, 1967).

Reproductive physiology of marsupials. G. B. Sharman. *Science* **167,** 1221 (1970).

Aspects of reproduction in the African elephant, *Loxodonta africana.* R. M. Laws. *Journal of Reproduction and Fertility*, suppl. 6, 193 (1969).

Male reproductive organs of the African elephant, *Loxodonta africana.* R. V. Short, T. Mann and M. F. Hay. *Journal of Reproduction and Fertility* **13,** 517 (1967).

Reproduction of elephant (*Loxodonta africana*) in the Luangwa Valley, Zambia. J. Hanks. *Journal of Reproduction and Fertility* **30,** 13 (1972).

The formation and function of the corpus luteum of the African elephant (*Loxodonta africana*). J. Hanks and R. V. Short. *Journal of Reproduction and Fertility* **29,** 79 (1972).

Persistence and parthenogenetic cleavage of tubal ova in the mare. C. H. van Niekerk and W. H. Gerneke. *Onderstpoort Journal of Veterinary Research* **33,** 195 (1966).

The origin of the equine endometrial cups. I. Production of PMSG by fetal trophoblast cells. W. R. Allen and R. M. Moor. *Journal of Reproduction and Fertility* **29,** 313 (1972).

Suggested further reading

Comparative Biology of Reproduction in Mammals. Ed. I. W. Rowlands. Symposium No. 15 of the Zoological Society of London (1966).

Biology of Reproduction in Mammals. Ed. I. W. Rowlands and J. S. Perry. *Journal of Reproduction and Fertility* suppl. 6. (1969).

Patterns of Mammalian Reproduction. S. A. Asdell. London; Constable (1965).

The Ovarian Cycle of Mammals. J. S. Perry. Edinburgh; Oliver and Boyd (1971).

33

2 Behavioural patterns
J. Herbert

Mating and the care of the young are crucial events in the reproductive life of any mammal.

An animal behaves in a particular way as the result of the interaction between its internal state and the stimuli it receives from the environment. When male and female rats are put together they are likely to mate, but may not if the male is very hungry or there is a predator nearby to be avoided, or if they have learnt that attempting to copulate in that particular environment will result in an unpleasant experience: for example, an electric shock. Furthermore, the hormonal secretions of both animals must also be sufficient to allow sexual behaviour to be evoked.

METHODS OF STUDY

There are various ways of studying the behaviour of animals. We must be able to measure their behaviour, and this often means making a judicious selection of behavioural components. If we are investigating sexual behaviour, the number of times the animals mount one another is likely to be important, but we may choose to ignore the number of times the animal blinks its eyes, grooms itself or explores its cage – unless we consider these activities to have a bearing on sexual interaction. Then, we have to devise ways of assessing levels of activity: if an animal ejaculates five times in an hour, we may be tempted to conclude that his sexual behaviour is at a higher level than another that ejaculates only once. An animal can be made to indicate the intensity of its behaviour by indirect means – requiring it to press a bar to gain access to the female, or cross a charged electric grid, in which case we can measure the current

34

it will tolerate before it stops trying to join its partner. Sometimes we refer to deductions made in this way as indicating variations in the animal's 'drive state', the physiological process that drives the animal towards a particular goal, in this case the opportunity to copulate.

We can study behaviour in surroundings resembling the animal's natural habitat. Here sexual and maternal activity are part of the animal's total behaviour, and we can observe the way these activities affect its way of life, and how the particular kind of society in which the individual lives influences the amount or direction of its reproductive behaviour. Thus, male deer live together peaceably for much of the year but split up and behave aggressively towards each other at the beginning of the breeding season in order to acquire and defend the territory necessary for their mating performance. A dominant male monkey is more likely than a subordinate one to mate with an attractive female, just as he is more likely to gain prior access to a choice piece of food. A female monkey may form a sexual consortship with a dominant male and she then becomes his equivalent in rank, but only for as long as their partnership endures.

Studying behaviour in such a setting is a very complex task. In order to define precisely the role of particular components, such as hormones or sense organs, we may put the animals in conditions in which the behaviour is likely to occur, but which exclude many other complicating factors; for example, we may pair a male and female for a standard time in an empty cage, and record what happens. This allows us to specify precisely the conditions of our experiment, and to repeat it at will. Although we can only obtain an incomplete knowledge of the total range of factors determining behaviour, these experimental methods are essential for more sophisticated analyses.

SEXUAL BEHAVIOUR

Sexual interaction can be divided into two parts, courtship and mating. The first includes all those patterns of behaviour by which the male and female signal to each other that they are in physiological readiness to copulate, and which arouse the sexual interest of potential partners. The most magnificent, complicated and stereotyped patterns are seen not in mammals, but in birds, fish and insects. In these groups, each animal has to follow a sequence of response and counter-response without flaw if mating is to take place; witness the spectacular display of the peacock, and the activity of the stickleback, who must go through an elaborate ritual before finally persuading the female to lay her eggs in his nest. But mammals, too, use specific signals to identify themselves to others as sexual partners. They may smell each other's face or genitalia, nibble or bite each other in a characteristic way, urinate over each other, or even engage in mock fights.

Mating behaviour in animals implies the acts of mounting, insertion of the penis (intromission) and pelvic thrusts by the male, and the adoption by the female of a characteristic hollow-backed posture ('lordosis') which facilitates his actions (see Fig. 2-1). Farmers can test whether sows are in heat by pressing on their backs; if they are, the females will respond by standing still and showing lordosis (Fig. 2-1). Not surprisingly, this response is more easily evoked by the boar than by the farmer! Female rats, guinea pigs and cats also will show lordosis if their backs are stroked when they are in oestrus.

During courtship, animals detect signals from one another by using their 'teloreceptors', the specialized sense organs of the eyes, ears, nose and skin that detect changes in the environment outside them.

Visual cues are very important. Females may behave in a particular way when near the male; rodents wiggle their ears and perform a curious darting run, and monkeys adopt a characteristic mating position. There may be changes in their appearance: in ferrets and mink, for example, the vulva swells enormously at oestrus, and several species of monkey exhibit marked swelling of the specialized skin that surrounds their genitalia. The male's appearance and behaviour also change:

antlers grown by male deer before the breeding season may serve to indicate to rival males that competition is likely. Some male monkeys, e.g. the mandrill, are brightly and distinctively coloured, and this colouration may become even more vivid during active breeding. Males usually become more aggressive towards each other during the breeding season.

Fig. 2-1. The mating postures of three female mammals, the monkey (*above*), the cat and the sow (*overleaf*). The monkey is posturing because a male (not shown) is nearby, the cat because she is being mounted by male; the sow is being tested for oestrus by a farmer.

Smell is equally important. This is interesting because smell is said to play a lesser part in the overall behaviour of an animal as its brain becomes more highly evolved. But in monkeys and maybe even in man, the sense of smell has an important place in sexual attraction. There is no doubt that many oestrous females give off a characteristic smell, which can be detected by adult males. An oestrous bitch, even if she is kept indoors, attracts dogs from the entire neighbourhood. Rams have difficulty in

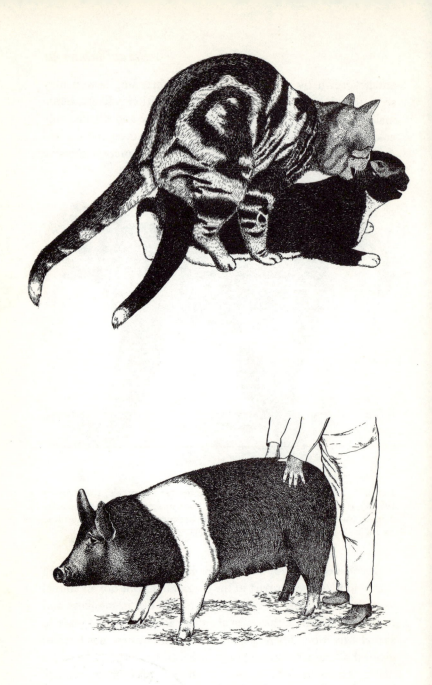

Fig. 2-1 (*continued*)

detecting which sheep in the flock are in oestrus if their sense of smell has been destroyed, and they can be deceived into treating an anoestrous ewe as if she were in oestrus by applying vaginal washings from another, oestrous, female to her genitalia. This last experiment (and many others) demonstrates that the sexually attractive odour comes from the female's genitalia.

Despite the traditional importance of the mating call in sexual folklore, auditory communication seems rather less important than vision and smell in the sexual activity of most mammals. Specific calls given by some males during the breeding season may attract females and warn off other males. Whilst we cannot define the role of these calls precisely, we may presume that they add to the general flow of information passing between male and female.

Experiments in which animals have been deprived of one or more of their senses have generally shown that no single sense is absolutely essential for mating activity and that the animal can compensate, at least to some degree, for loss of one faculty. Usually, the more restricted the sensory information available, the less the likelihood of sexual interaction, though male hamsters do not mate at all if deprived of the sense of smell.

Effects of castration and hormone administration in non-primates

Females. The fact that reproductive activity in females is cyclic is dealt with in greater detail in Books 1 and 3. Females of many non-primate species are only in 'heat' for a limited time during their oestrous cycle; at other times, they will usually refuse to mate. Removal of the ovaries promptly and completely abolishes all sexual activity (see Fig. 2-2). Put a spayed female cat into a pen with a male: if he attempts to mount or even approach her, her response is the aggressive display that cats are so admirably equipped to produce. But give the same female a single injection of oestrogen, and her response changes dramatically; instead of attacking the male, she rubs against him, crouches in front of him, allows him to seize her by the scruff of the neck, elevates her pelvis and makes characteristic stepping movements with her hind legs (Fig. 2-1). Such experiments clearly indicate that sexual activity in these females is dependent upon the presence of ovarian hormones, which are secreted in such a way that females become active sexually at about the time when ovulation, and hence, fertilization, is most likely to occur.

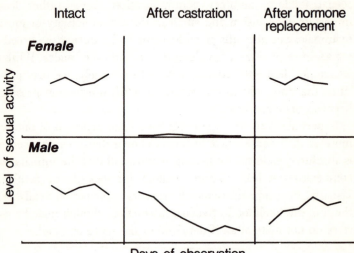

Fig. 2-2. The effects of castration and subsequent hormone treatment on sexual behaviour in non-primates. The female shows very rapid loss of sexual activity after being castrated, and an equally abrupt restoration after being given hormones. The male shows a much more gradual change both after castration and subsequently following treatment with hormones.

The relative importance of the main ovarian hormones, oestrogen and progesterone, in eliciting mating behaviour varies in different species. Female carnivores (cats, ferrets, dogs) can be brought into heat with oestrogen alone. In female rodents (rats, mice, guinea pigs and hamsters) heat is more easily induced by oestrogen followed, several hours later, by a larger amount of progesterone. Sheep require oestrogen preceded by progesterone.

The dose of the two hormones and the time interval separating their administration is quite critical, but we do not yet understand exactly how the two hormones interact upon the brain, or why there should be the species differences just described. It seems that progesterone secretion may also bring oestrus to an end in some species. In the rabbit and ferret the very act of mating induces ovulation and corpus luteum formation; the latter in turn terminates oestrus behaviour, because it secretes progesterone.

Males. There is some evidence that rather indefinite changes, roughly comparable to the oestrous cycle of the female, may occur in the males of some species but they are much less pronounced. Castrating a male causes his sexual activity to decline but much more gradually than in the female (Fig. 2-2); male dogs may take months or even years to reach their minimal level. Even when this point is reached, males may continue to show some degree of sexual activity. What remains depends upon several factors, but particularly on the age of the animal at castration; prepubertal castration is more effective than postpubertal. The sexual experience the animal has had before and after the operation is also important; male cats allowed to mate regularly before castration showed more residual sexual behaviour afterwards than cats of comparable age that had been kept apart from females. Male rats, castrated and then allowed frequent opportunity to mate, continued trying to do so for longer than animals given access to females at only infrequent intervals after the operation. Different components of the male's sexual pattern disappear at different rates: first the ability to ejaculate and then the ability to achieve intromission, but the male may still continue to mount an oestrous female for some time.

Giving a castrated male testosterone restores him to full sexual vigour; the level of his activity depends not upon the amount of hormone but upon neural factors on which the hormone acts. For example, two strains of guineapigs are distinguished by marked differences in their levels of sexual activity; giving both strains an excess of testosterone after castration results in the reappearance of their characteristic preoperative differences, and the lower-activity strain cannot be converted to the higher-activity pattern by giving still more hormone. We shall consider what we know about the determination of different levels of sexual activity later in this chapter.

Masculinity and femininity

We have spoken of mounting, intromission and ejaculation as the 'male pattern' and of lordosis or its equivalent as the 'female pattern' of behaviour. However, during the course of normal sexual interaction, each animal may display patterns more usually associated with the opposite sex. For example, female cows or cats in heat frequently mount one another in the male manner. Such behaviour is termed 'heterotypical', as opposed to the usual 'homotypical' pattern. Similarly, a male may mount another male either because he is responding to a 'male' stimulus in an unusual manner, or because the mounted male is behaving in some way like a female and thus eliciting the normal response to a real female. In a behavioural context, we cannot define maleness and femaleness according to the genetic sex of the animal, but only according to the pattern of behaviour. If a potent male and an oestrous female are paired together, the male is likely to mount the female because of the stimuli given by the one towards the other and the internal state of both. However, in some circumstances, the female may mount the male, or they may fight, or show mutual disinterest. The outcome can never be predicted with absolute certainty; we can only infer that the probability of a given response is particularly high. It is thus of the greatest importance to specify not only the behavioural pattern, but the object towards which it is directed. Nevertheless a male, confronted with an oestrous female, is more likely to behave in the homotypical than the heterotypical manner. Why?

Giving ovarian hormones to a castrated male, or testosterone to an ovariectomized female, is not very successful in transforming their behaviour to that of the opposite sex; in fact, the 'wrong' hormone may be almost as effective as the natural one in inducing homotypical behaviour. Something other than the adult male's hormones, therefore, determines his maleness. The answer might seem to be the male (Y) chromosome (see Book 2, Chapter 2), but this is quite incorrect. We know of instances in rats, mice and man in which genetic males with a normal XY chromosome constitution look and behave just like females because of a genetic defect that makes them unresponsive to testosterone; the condition is called 'testicular feminization'. We can also convert genetic male rats to behavioural females by castrating them on the first day of life. Conversely, if a newborn female rat is given a single dose of testosterone, she will behave and respond more like a male when she grows up. The same general features apply to other aspects of sexual differentiation (see Book 2, Chapter 2). Clearly, then, the presence of androgen at a critical point in the animal's life produces some permanent change which determines that the animal will behave as a male when it is adult. In the absence of this neonatal stimulus, the animal will behave as a female. We do not know what this change is, although alterations in the content of RNA in certain brain cells have been suggested. Animals born in a more mature condition than rats, such as guinea pigs, monkeys and probably man, pass through this critical phase while still in the uterus.

Femaleness, therefore, is suppressed by androgens early in life. But whether maleness is induced solely by androgens acting on the brain is still not clear; absence of androgen in early life also affects the growth of the penis, and without a normal penis an animal cannot show the complete range of normal male sexual behaviour.

Behavioural patterns

People have been tempted to postulate the existence of separate 'male' and 'female' centres in the brain but there is no satisfactory proof that they exist. However, spaying a female promptly obliterates all homotypical behaviour whereas her heterotypical mounting behaviour takes much longer to disappear, just as it does in a castrated male. This suggests that the mechanisms concerned in male and female types of behaviour are rather different.

The brain and sexual behaviour

During sexual behaviour many parts of the brain are active. For example, the visual and olfactory pathways are receiving and analysing the signals sent out during courtship, and the pathways controlling movement allow the animals to take up the characteristic postures necessary for mating. These systems are not specifically concerned with sexual interaction. At the next moment, the animal may be occupied in building a nest, attacking another animal, or eating food, and the same pathways are involved, though the information they carry and the pattern of activity they produce may be very different. But, as we have seen, giving an animal gonadal hormones produces marked and specific effects on sexual behaviour and a natural question to ask is whether these effects are brought about by an action on a specific part of the brain.

If a small area of the brain called the hypothalamus, lying just above the pituitary gland, is destroyed we cannot induce sexual behaviour by treating the animal with hormones. Since the hypothalamus is also involved in many other processes, such as eating, drinking, regulation of blood pressure and body temperature, aggressive behaviour and the function of the pituitary, we have to be sure that the failure of the animal to mate is not simply because its well-being is disturbed by the operation. To produce the converse effect by implanting a minute amount of hormone directly into the hypothalamus is more convincing (see Fig. 2-3). Androgen implanted in this way into castrated males restores their sexual performance; oestrogen put into spayed females causes them to come into oestrus. The same treatments to other parts of the brain are ineffective. The action of hormones in eliciting sexual behaviour therefore seems to take place at least partly in the hypothalamus. This is not to say that other parts of the nervous system cannot respond to hormones; there is evidence that testosterone, for example, can act directly

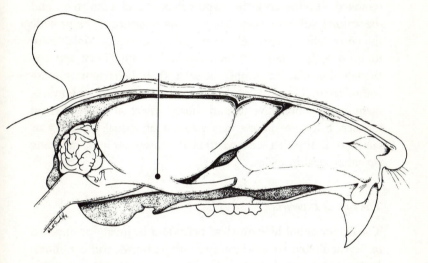

Fig. 2-3. A section through the head of a rat, showing a small pellet of hormone implanted directly into the hypothalamus at the base of the brain. The hormone is on the end of a needle which is fixed to the skull.

on the spinal cord to alter certain reflexes concerned in sexual behaviour. Damage to other parts of the brain, particularly the so-called 'limbic system' can also result in changes in sexual activity, but other categories of behaviour may alter as well.

These results also suggest that the organizing effect of andro-gen on sexual behaviour in early life occurs in the hypothalamus. In adulthood, gonadal hormones seem to act by evoking what-ever patterns of behaviour may have been laid down in the brain, particularly in the hypothalamus, by hormones acting during development. But the way this comes about is still not understood. All we know is that hormones can directly affect electrical activity of the brain cells in the hypothalamus, and that a discrete part of the hypothalamus, distinct from those areas involved in other regulatory functions, seems to be particularly concerned with sexual activity.

A large amount of the brain of a female rat or cat can be removed, leaving only the hypothalamus and brainstem, and the animal will still show the lordosis response if we provide the correct hormones and the necessary stimulus. Males seem to need rather more of their brain for copulatory activity, though this may be simply because their behaviour is more complicated. Thus, removal of the areas concerned with movement has a more obvious and drastic effect in males than in females. Such operations may also cut off visual, olfactory or tactile stimuli which seem particularly important in maintaining sexual arousal in males.

The role of experience

The debate about how much of behaviour is 'innate' as opposed to 'learned', and what we mean by these terms, still continues. Let us simply say that any behaviour, however much it depends upon genetically determined components within the brain, can be modified by experience, and that what an animal learns depends not only upon its experience but upon the kind of brain it brings to the task.

There is little evidence that the experience of mating has a major role in the reproductive activity of female non-primates; they are as likely to show the complete pattern of female behaviour on the first occasion they mate as on subsequent ones. However, a male's behaviour, and even male-like behaviour in females, is more susceptible to the effects of experience. Female guinea pigs reared in isolation display less of this kind of behaviour than their socially reared contemporaries. Males reared alone, or denied tactile contact with other animals, are inferior in sexual performance when they grow up. They may take a long time to recover, if they ever do. It is wise to recall that rearing animals in isolation is a highly abnormal procedure and that this may result in widespread behavioural aberrations of which sexual inadequacy is only one.

If an animal experiences pain or fright in a particular situation, it may eventually lose its sexual responsiveness in the same surroundings. Male cats attacked repeatedly by females they are about to mount gradually lose sexual interest; but if we put them into new surroundings, and allow them to acclimatize, their sexual activity returns. Conversely, we can increase the possibility of the animal showing sexual activity by habituating it to conditions in which successful copulation has been the rule.

Sexual behaviour in primates

How far does the information we have been considering apply to the primates, the group that includes monkeys, apes and man? Has the primate brain evolved out of the control exercised by hormones in non-primates?

Female. The female primate differs from the females of other mammals by not showing a sharply circumscribed period of heat during the reproductive cycle. This is not to say that variations in sexual activity do not occur; in general, activity of both monkeys and women tends to be greatest during the middle of the menstrual cycle, with a second peak just before menstruation (see Fig. 2-4). Removal of the ovaries from female monkeys decreases, but does not abolish, sexual interaction with the male. Ovariectomy in women is reported to have little effect on libido, although there are obvious problems about measuring it. Does this mean that hormones have little or no role in primate sexual behaviour?

If a female monkey is spayed, the sexual activity of her male partner decreases. Her own receptivity, or willingness to mate, remains little altered, but she has great difficulty in attracting him. We can restore her sexual attractiveness to the male by applying a small amount of oestrogen directly into her vagina, or by giving her a larger dose by systemic injection. Moreover, if a male is made anosmic (unable to smell) he is less easily

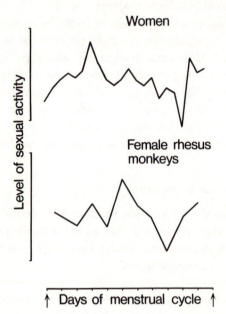

Fig. 2-4. Variation in sexual activity during the menstrual cycle (menstruation shown by arrows) in women and female rhesus monkeys. There are peaks at mid cycle and just before menstruation, whereas lower values occur during the luteal phase. (Data from J. R. Udry and N. M. Morris. *Nature, Lond.* **220,** 593 (1968); and J. Herbert. In *Progress in Primatology*, p. 230. Ed. D. Starck. Gustav Fischer (1967).)

aroused by the oestrogen-treated female. These findings suggest that the ovarian hormones affect a female's sexual attractiveness by altering her vaginal odour. We all know how frequent the use of perfume is in our own society as a sexual attractant; does this represent a socially modified version of the same mechanism? In some species (e.g. baboons, chimpanzees) the skin around the female's genitalia shows enormous swelling at mid-cycle but we cannot yet say whether this acts as an added visual stimulus concerned in sexual attractiveness.

We also know that the female monkey's receptivity is controlled by androgens, rather than oestrogens; in the female these come mainly from the adrenal gland, although a little comes from the ovary. A female monkey given androgen tries to make the male mate with her more frequently. Conversely, if both her adrenals and ovaries are removed, and she is given oestrogen so that she is sexually attractive, she will refuse to allow the male to mount her; she is restored to normal by being treated with small amounts of androgens. Therefore, while the ovarian hormones control the female's attractiveness, the adrenal (and ovarian) androgens control her receptivity, presumably by acting upon some part of her brain. The secretions of the adrenal glands do not show such pronounced cyclical changes as those of the ovaries, which may explain why female primates are not subject to circumscribed periods of 'heat'.

So far as we can tell at present, the same mechanisms may operate in women. Giving women testosterone or other androgens increases their sexual desire, whereas removing their adrenals results in loss of much of their libido. Women taking the contraceptive pill, which is usually a mixture of an oestrogen and a progestagen, may complain of low sexual 'drive'. Interestingly enough, such women excrete small amounts of androgens in their urine; we do not yet know whether giving them low doses of androgens would restore them to normal sexual vigour. So in primates it seems that hormones continue to exercise an appreciable control over sexual activity in the female.

Male. There is little evidence about the effects of castration and hormone treatment in male primates, though the general opinion is that sexual activity declines after operation. Prepubertal castration in man, as in animals, is much more effective than post-pubertal castration, and the effects can be reversed by treatment with testosterone. Recently, the anti-androgenic substance cyproterone has been reported to diminish sexual drive in humans. This seems odd, because it has no effect on male rats or hamsters although it can apparently suppress the mating activity of boars. In general, the results of castration in man and monkeys are variable, and male libido takes a long while to decline. The decreased sexual potency of many older men is not caused by decreased testosterone levels, and does not usually respond to treatment with hormones.

Differentiation of sexual behaviour in primates

About 20 years ago the vogue was to give pregnant women androgenic steroids early in gestation in an attempt to prevent threatened abortions. As a result, many female children were born with their external genitalia 'masculinized' to varying degrees; in fact, some resembled boys so much that they were brought up as males (Fig. 2-5). At some point in their childhood the mistake was recognized and the children were then reared as girls. Not surprisingly, these children had difficulty in adjusting to their true sex; in general, when the change was made before about 2 years they adopted their new sexual role more successfully than if they were reared as boys for longer periods. This finding gave rise to a theory of the development of sexual behaviour in humans which differed radically from that derived from work on rodents described above. According to this theory, children are born sexually neutral and grow up to look upon themselves and behave as either boy or girl depending on the sex their parents have ascribed to them ('assigned sex'). Even children with external genitalia contrary to their assigned sex were found to accept their sexual roles and behave accordingly. The theory therefore gives particular weight to the influence of the parents' attitudes upon the child's sexual role, and very little to the influence of prenatal hormones.

The only experimental tests of the development of sexual behaviour in primates are those in which pregnant rhesus monkeys were given androgen. Their female offspring were masculinized, as in human beings, and as they grew up they showed play behaviour and infantile sexual behaviour more characteristic of males than females. We do not yet know how

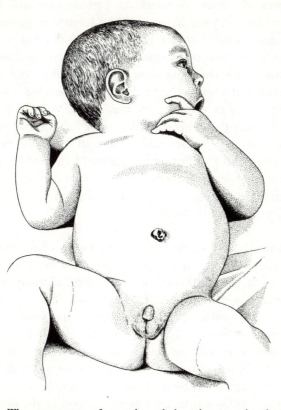

Fig. 2-5. The appearance of a newborn baby whose mother had been treated with androgens during pregnancy. The baby is, in fact, a genetic female. (From J. L. Hampson and J. G. Hampson. In *Sex and Internal Secretions*, p. 1405, Fig. 23-1. Ed. W. C. Young. © 1961 The Williams & Wilkins Company, Baltimore.)

they will behave sexually as adults, though they seem to develop male-like mounting behaviour as they grow up, nor can the evidence as it now stands tell us whether the effects are due to processes analogous to those described in rodents, or whether their altered appearance caused a significant change in mother–infant relationships.

In man, two mechanisms may be at work: the effects of androgens on innate behavioural patterns overlaid by the effects of assigned sex of rearing. The amount of genital masculinization induced by giving androgen to the mother might have been paralleled by masculinization of behavioural mechanisms, and thus there might seem to be an approximate correlation between the way a child is brought up and the amount of difficulty it has in adopting its 'true' sexual role. This by no means negates the general findings that behavioural patterns are more easy to establish early in life than later on in development. The position is further complicated by recent knowledge that the receptors responsible for the action of androgen in peripheral structures, such as the genitalia, differ in some way from those present in the brain.

Homosexuality. Sexual activity with a member of the same sex, or the adoption of patterns more appropriate to the other sex, is part of the normal behavioural repertoire of many animals, and the same applies to man. There is a spectrum ranging from people who are exclusively heterosexual to those who are principally homosexual, though at exactly which point we should label someone 'homosexual' is arbitrary. We must also recognize that the term homosexual includes a number of subdivisions, though they are not distinct: those who tend to behave in a way appropriate to their own sex but with others of the same sex, those who regard themselves and wish to become members of the opposite sex, and those who may derive sexual satisfaction from dressing as members of the opposite sex (the last group are not necessarily, or even usually, homosexual). The amount that homosexual behaviour contributes to total sexual activity can vary considerably at different periods of life. Moreover, even within a given category of homosexual activity there are in fact as many variations as there are in heterosexual activity. Even the traditional division of homosexuals into 'active' and 'passive' (or 'butch' and 'femme') is not always a distinct one in practice.

Naturally, numerous explanations have been advanced as to why a person should be homosexual. Repeated attempts to prove abnormal hormone secretion have failed, and hormone treatment of homosexuals is ineffective in altering the direction of their sexual interests. If one identical twin is homosexual, the other is very likely to be homosexual, but this relationship is not true of non-identical twins of the same sex. This has been interpreted to mean that some genetic factor might be implicated, although the validity of the original observations has since been criticized. There is no evidence that the sex chromosomes in homosexuals are in any way abnormal, nor do aberrations in these chromosomes necessarily result in deviant sexual behaviour (see Book 2, Chapter 2). Psychoanalysts have suggested that parental attitudes may be implicated; in particular a hostile or indifferent father or an oversolicitous mother may evoke homosexual tendencies in children, whilst inhibiting heterosexual ones. Many societies bring strong pressure to bear against homosexual behaviour, and this may in itself influence the direction of a person's sexuality, and induce behavioural abnormalities in those who, despite such constraints, continue to act homosexually.

Bearing in mind the factors controlling the development of sexual behaviour in rodents, it is not surprising that similar mechanisms have been postulated to cause homosexuality in humans. Abnormally high amounts of androgens in embryonic life could perhaps induce lesbianism in females, and low levels might result in homosexual males. Such mechanisms may play a part, though there is still no convincing evidence that they do. Even so they could not account for the variations in homosexual behaviour described above. For example, it is difficult to see how one could produce in this way a male homosexual who preferred to treat other males as females, and a male who, whilst regarding himself as male, usually adopted a feminine role towards other males.

One reason for our limited knowledge of homosexuals is that

it is based upon people who for various reasons seek psychiatric help. We are therefore still very ignorant of the incidence and nature of homosexual behaviour in 'normal' society. We also have the problem, common to all studies of human sexuality, of not being able to measure confidently the degree and nature of homosexual behaviour.

MATERNAL BEHAVIOUR

Variations occur in social organization and physiological adaptations in mammals, but there is a constant demand upon the mother to provide warmth, food and protection for the young. The amount of maternal behaviour required by the young depends upon the condition in which they are born. Some are born very immature, 'altricial'; for instance, kittens are blind and deaf at birth, and rats are born naked and feeble. On the other hand in some species, particularly those adapted to living in herds, the young are born much more mature, 'precocial', and are capable of a considerable degree of independence immediately after birth. Although there is great variation between species, there are four principal ways by which mammals provide suitable environments for their young:

1. The infants are born extremely immature and are carried by the mother in a specialized pouch which, in some ways, acts as a kind of external uterus (marsupials).

2. The mother builds a nest in which the immature young are kept warm (rodents, rabbits, bears).

3. The young are carried by the mother for most of their early life. This provides not only warmth from her body, but keeps the infant near its food supply, protects it from predators and allows it to move with its mother as the rest of the social group moves. This is a method adopted principally by primates and by bats (Fig. 2-6).

4. The young are born very mature and are capable of fending for themselves with only a minimal amount of maternal care,

mainly provision of milk (deer, antelope, cattle, sheep and horses).

There are some very unusual patterns of maternal behaviour; one is shown by females of certain species of tree shrews, which

Fig. 2-6. A mother langur cradling her infant. This is the characteristic simian position for caring for the very young. (From P. Jay. In *Maternal Behaviour in Mammals*, p. 282, Fig. 1. Ed. H. Rheingold. © 1963 John Wiley & Sons, Inc. New York.)

are usually (but probably incorrectly) classed as primitive primates. They actually segregate their young in separate nests from those that they occupy themselves and only visit them every 2 days or so to feed them. The young of these species clearly need to possess extraordinary ingestive and digestive capabilities.

Characteristics of maternal behaviour

Although true maternal behaviour starts with the birth of the infant, preparation for it occurs during the later stages of pregnancy in many animals. The female may move away from her fellows and begin to build a nest. The chest fur of female rabbits becomes loosened at this time so that it can be more

Fig. 2-7. A cat licking her newborn young to remove the fetal membranes. (From T. C. Schneirla, J. S. Rosenblatt and E. Tobach. In *Maternal Behaviour in Mammals*, p. 134, Fig. 5. Ed. H. Rheingold. © 1963 John Wiley & Sons, Inc. New York.)

easily plucked out to line the nest. Experiments have shown that a female's maternal response to young animals increases gradually as pregnancy advances, though it only attains really high levels shortly after birth.

As the young are born, the mother licks them to remove their fetal membranes (see Fig. 2-7) and then she eats the placentae (this occurs even in herbivorous species). We do not know exactly what prevents her from going on to eat her young, though this probably has something to do with the way they move and the noise they emit. The infants are certainly potentially edible, for they are frequently eaten if something goes wrong with the mother's maternal responses, or if the infants are deformed, or under conditions of grave danger or serious crowding.

Mothers of many species show retrieving behaviour; that is, they will pick up, or guide back, infants that stray too far away from them or from the nest. Infant monkeys are allowed to play away from their mother but are quickly clutched to her chest if danger threatens, and we are familiar with a similar response from the human mother in moments of crisis. Some precocial young, however, are left by themselves, and their mothers may even move away from them if danger threatens, presumably to divert attention from them. The young goat is treated in this way; the young lamb, however, trots after the ewe come what may. Some workers have suggested that the relative independent-mindedness of the adult goat in contrast to the well-documented conformity of sheep stems from these differences in their experience during early life.

The young animal seems to treat its mother as a source not only of comfort and nutrition, but also of security. Infant monkeys placed in a strange environment will explore it and show much less stress if their mothers, or even inanimate models that have acted as mothers, are present as well. Exactly the same results have been obtained in studies on human beings. As it grows older, the infant becomes more independent. This is a two-way process involving both the mother and the infant; the infant's maturing muscular and nervous system allows it greater freedom, whilst the mother gradually becomes less and less tolerant of its activities. Thus, as time goes on, a subtle change in the relationship between mother and infant occurs; whereas shortly after birth the mother will initiate most of the contact between herself and her infant (see Fig. 2-8), gradually the emphasis changes and the infant has to take a greater part in maintaining the relationship if it is to endure (Fig. 2-9). Eventually, the bonds are broken by the mother becoming more and more unresponsive to the infant and actively discouraging its attempts to continue being nursed or remaining in contact with her. The infant, on its part, is spending more and more of its time in contact with other members of its society and its environment.

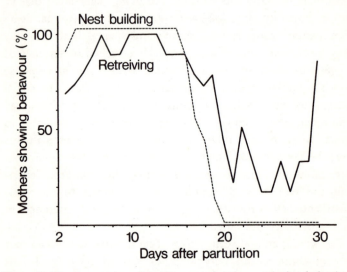

Fig. 2-8. The proportion of mother rats showing retrieving behaviour (continuous line) or nest building (pecked line) following the birth of their litters. Note how both measures decline about 15 days after parturition, retrieving behaviour more gradually than nest building. (After J. S. Rosenblatt and D. S. Lehrman. In *Maternal Behaviour in Mammals*, p. 30, Fig. 6. Ed. H. Rheingold. © 1963 John Wiley & Sons, Inc. New York.)

Communication between mother and young

We have already seen how the mother characteristically licks her newborn young, and she spends a good deal of her time doing so during the early part of infancy. Indeed, females often show an increased tendency to lick anything, particularly themselves, during pregnancy. The smell or taste of the young may be important in maintaining and reinforcing her maternal responses towards them, though females prevented from licking their young, or eating the placenta, can still function as good mothers. Licking also has an important role in some species in that it stimulates the young to defaecate and urinate. The mother may swallow these excreta, and so avoid contamination of the nest.

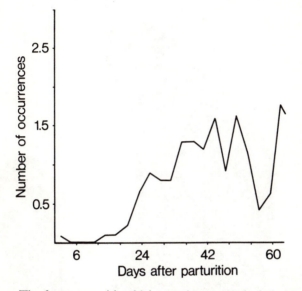

Fig. 2-9. The frequency with which puppies approach their mother. They begin actively to initiate contact with her increasingly often as they grow older. (After H. Rheingold. In *Maternal Behaviour in Mammals*, p. 185, Fig. 8. Ed. H. Rheingold. © 1963 John Wiley & Sons, Inc. New York.)

Many young animals emit a special call if they are separated from their mother, and this allows her to locate them. These calls may be surprisingly loud, or ultrasonic as in the case of mice. The human mother can distinguish various qualities of crying by her baby; if one plays a recording of a baby crying because it was hungry she is likely to respond in a fairly leisurely manner; but her response to one made whilst the baby was in pain is immediate and anxious.

Tactile sensations are very important, particularly for the infant. The infants of many species will cling to anything that is soft, furry and warm. Infant monkeys given a choice between artificial mothers that feed them, but are made of hard material, and those made of soft material, but that do not provide food, chose the latter. The infant uses its tactile information to locate

Fig. 2-10. Mother and young rhesus monkey. (From H. F. Harlow, M. K. Harlow and E. W. Hansen. In *Maternal Behaviour in Mammals*, p. 254, Fig. 15. Ed. H. Rheingold. © 1963 John Wiley & Sons, Inc. New York.)

the mother's nipple by using the so-called 'rooting' reflex, though she may give it some help. You can demonstrate this reflex in the newborn human baby by gently touching it near its mouth with your finger, when it will turn its head and try to take your finger in its mouth as if it were a nipple.

The appearance of the young seems to be important in eliciting maternal responses, though it is difficult to be precise about the exact features that are of significance (see Fig. 2-10). Some young animals have a special colouration; young rats are naked and pink in contrast to the white or black or brown adult, infant baboons have a much darker coat than the greenish-yellow or brown adult. The fat cheeks of the human baby assist suckling and they may also act as a stimulus tending to elicit maternal behaviour.

The development of smiling in human infants has been much studied, for this is an important way by which the infant communicates and establishes contact with its mother. Visual communication begins with the infant becoming able to fixate its mother with its eyes. Newborn babies cannot do this because their nervous system is so undeveloped. After a few months the infant begins to smile, at first to itself, then at any human face and finally, more specifically, at its mother's face. At this point it may begin to react to strange faces with fear or wariness. But being able to see its mother is not essential for smiling to develop, for blind babies begin to smile in the direction of their mother's voice, though their smiles may be rather more abbreviated than normal.

Mothers and young therefore recognize each other. The young can also recognize their 'mother' even if she is from a different species, for young goats reared with a particular bitch recognize and prefer her to other bitches. Getting a mother to accept and rear strange young is sometimes very difficult, and to do so generally requires some stratagem (e.g. preventing her smelling them) to persuade the mother that the strangers are, in fact, actually a part of her own litter.

Hormones and maternal behaviour

The non-pregnant female behaves in a very different way from the pregnant or lactating one towards young animals. Since it is well known that the placenta or the ovaries usually produce large amounts of oestrogen and progesterone during pregnancy and that shortly after birth the pituitary gland begins to secrete prolactin (see Book 3), people have naturally wondered how far these substances are responsible for the characteristic changes in a female's behaviour that accompany the birth of her young.

Most of the research on this topic has been conducted on rodents, and has involved injecting a non-pregnant animal with various doses and combinations of the three hormones, measuring her behaviour towards very young animals, and comparing it with that of a normal mother. These attempts have been rather unsatisfactory in most cases. Maternal behaviour has usually proved possible to induce in only a small number of treated animals, whereas nearly all normal females show such behaviour after giving birth, so that the experimental treatment has not really duplicated the natural condition. However, not only may the levels of the three hormones be important, but also the way in which they are secreted. Progesterone secreted during pregnancy may condition the animal to oestrogen and prolactin, but prevent the premature appearance of maternal behaviour. After birth, progesterone levels decrease sharply and this may facilitate the appearance of maternal responses under the combined influence of the other two hormones. Some think that the hormonal basis of maternal behaviour is very different from that of lactation, and in particular, the role of prolactin, once thought to be *the* maternal hormone, is still not accepted by everybody. Maternal behaviour can occur in the absence of reproductive hormones, and some experimenters even found that removing the pituitary gland was an excellent way of inducing rats to behave in a maternal fashion! Once maternal behaviour has been initiated naturally, it is easier to maintain with hormones than to induce in non-pregnant animals.

Maternal behaviour is also much harder to induce in males than in females, though species and strains differ. People have wondered whether this might be due to the same imprinting effect of neonatal androgen that produces the differences in sexual behaviour. But a recent experiment in which male hamsters were castrated early in life did not result in their showing maternal behaviour towards young in the same way that a female would.

Role of the young

If a non-pregnant female rat is given a litter of baby rats, she may ignore them, or attack and eat them, but she seldom shows any signs of behaving maternally towards them. If she is presented with successive litters of pups, she gradually begins to mother them and if this treatment continues for several days she will eventually develop a high degree of maternal behaviour. Evidently the presence of young has produced a gradual change within the female which results in her beginning to respond to infants in a maternal rather than an aggressive or disinterested way. Since this effect can be observed in ovariectomized or hypophysectomized females, it does not seem to be dependent upon hormonal changes within the female. If only we could specify the underlying changes in the female's brain, we should know a great deal more about the physiological basis of maternal behaviour.

The ability of young to induce maternal behaviour depends upon their age, very young infants being more effective than older ones. We can prolong a female's maternal responsiveness for a time by substituting newborn pups from other litters for her own growing young. Continued maternal responses also depend upon the young being present. If we remove, for a time, a female rat's young shortly after birth, she rapidly loses the ability to respond to her litter when they are returned, whereas if the young are left with her a few days, she tolerates their temporary removal much better. Thus there seems to be a period, shortly after birth, which is particularly important in establishing maternal behaviour. This is not the birth process itself, since young delivered by Caesarian section provoke normal responses by the mother. Neither does the effectiveness of the young depend upon their ability to elicit lactation by suckling, even though the hormones concerned in milk secretion and maternal behaviour seem very similar.

Role of experience

Many zoo keepers expect that females are likely to have difficulty in rearing their first litters. It has even been suggested that first litters have no survival value on their own, but act only as a rehearsal for subsequent ones which will form the female's contribution to the propagation of her species. Not all experimental evidence agrees with this, but an experienced mother is generally more relaxed and capable than a naive one, and is more likely to rear her babies without neglecting them, preventing them feeding, eating them or accidentally killing them. Female monkeys seem very gauche and nonplussed by their first infants, and they may hold them in such a way that the infant has difficulty in reaching the nipple.

That the role of experience and of neonatal hormone treatment have such different effects on a female's sexual and maternal behaviour is interesting. Although we are accustomed to thinking of both categories of behaviour as part of the 'feminine' constitution, they may in fact be different mechanisms. One example of this, from our own society, is that lesbian women who may have little or no sexual interest in men nevertheless may greatly desire to have children and may become excellent mothers.

The young in society

It would be a mistake to think that all the important reactions in an infant's life occur between it and its mother. Other factors, which have been particularly carefully studied in monkeys and apes, have great significance.

The young monkey seems to be a focus of interest for all the females in his group. They try to touch and play with him, though his mother may be reluctant to allow this to happen.

Nevertheless, one particular female may form a specially close relationship with the mother and her infant and is known colloquially as an 'aunt', though this term does not imply a blood relationship. Aunts may take an active part in the rearing process and their presence has a definite effect on the behaviour of the young monkey. Similar factors may be a feature of cultural differences in our own species: in the United States, it has been shown that Negro children in Washington are more likely than Caucasian infants to be in the company of people other than their parents.

Parental behaviour by males is said to occur in some Japanese monkeys, but, in general, males seem relatively indifferent to infants, though they may respond to their cries of distress. An important point is their tolerance towards the infant's activities. A baby monkey may play with, climb over and have mock fights with a huge and dominant male monkey with complete impunity, activities that would mean severe injury or even death should they be attempted by more adult members of the troop. As they grow up, infants lose this immunity. Young males are particularly liable to be attacked by adult males and in some species may be driven from the group, either to join another or to live a solitary life, or to form a bachelor band with other similarly dispossessed young males.

If an infant monkey is removed at an early age from its mother, and is brought up by itself or with only an artificial 'mother' for company (see Fig. 2-11) it grows up to be very abnormal. It refuses to play and sits in the corner of its cage with its face buried in its arms. When it grows up it will have great difficulty in behaving normally in social contact with other monkeys and it will usually refuse to mate; these effects can be largely prevented if it is allowed to play with other infants when it is young. Even baby monkeys that are separated from their mother for only a few days shows changes in behaviour that are still detectable more than a year later. Such experiments have the greatest interest in view of our attempts to account for delinquent or antisocial children in our own society, a high

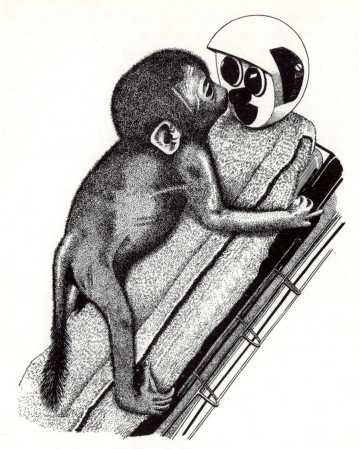

Fig. 2-11. An infant monkey clinging to an artificial mother. (After H. F. Harlow. In *Frontiers of Psychological Research*, p. 93. Readings from *Sci. Amer.* selected by S. Coopersmith. Freeman, San Francisco and London (1966). This figure © Gordon Coster, Chicago.)

proportion of whom seem to have lacked adequate parental care.

Other, more subtle qualities may affect the infant's development. Baby rats handled by an experimenter during early life grow up to show less reaction to an unfamiliar environment than their unhandled contemporaries, and young lambs can be tamed by being handled ('gentled') repeatedly. But the same effect can be observed if only the mother is handled; presumably she transmits this experience in some way to her infants. A remarkable finding in monkeys, and one reminiscent of the hereditary principles operating in our own society, is that the young of dominant female monkeys are themselves likely to be dominant over the infants of subordinate females as they grow up. We do not yet know how far this is attributable to the experience of being reared by a dominant mother, or to genetic qualities transmitted by her.

Although the process of rejection and the gradual assumption of the adult way of life might seem to signal the end of the mother–infant relationship, this is not necessarily the case. If we study very carefully the groups that gather to groom each other, we find that infants, even when they are grown up, tend to groom their mothers and siblings more than other monkeys. Chimpanzees continue to associate with their mothers several years after they attain adulthood. Now, one of the most consistent findings in comparative anthropological studies is the 'incest' taboo, the prohibition of mating between mother and son and other close relatives. Recently the same phenomenon has been reported to exist in the Japanese macaque monkey and in chimpanzees, in which adult males seem very unlikely to mate with their mothers. If this is true, it offers a new insight into the biological origin of the taboo in human society.

Although we have been considering side by side studies on monkey and human babies, there are clearly differences between the two based upon social, neurological and cultural factors. One such difference which may or may not be important is that young monkeys spend nearly all their early life clinging to their mother. In man, a species that has lost most of its fur, clinging is very much reduced and may be actively discouraged in some societies. This has led one eminent worker to suggest that a crucial test of this species difference would be to persuade a human mother to rear her infant while she continually wears a fur coat!

Behavioural patterns

The study of behaviour has long been a fascinating pastime, but something more in the nature of a hobby than a serious pursuit. In recent years attitudes have changed, and behavioural psychology has now plainly become an exact science, with important contributions to make on the functions of the nervous and endocrine systems, and the effects of environmental agents. More than that, we are coming to understand better the nature of our own motivation and activity, and to appreciate some of the underlying forces that shape our society.

SUGGESTED FURTHER READING

The hormones and mating behaviour. W. C. Young. In *Sex and Internal Secretions*, 3rd edition, vol. 2. Ed. W. C. Young. Baltimore; Williams & Wilkins (1961).

Sex hormones and other variables in human eroticism. J. Money. *Ibid.*

Reproductive behaviour. Ed. J. S. Perry. *Journal of Reproduction and Fertility* suppl. 11 (1970).

Abnormal sexual behaviour. J. H. J. Bancroft, F. E. Kenyon and J. Randell. *British Journal of Hospital Medicine* **3**, 168–214 (1970).

The study of mother–infant interaction in captive group-living rhesus monkeys. R. A. Hinde and Y. Spencer-Booth. *Proceedings of the Royal Society*. B **169**, 177–201 (1968).

Maternal behaviour in rodents and lagomorphs. M. P. M. Richards. In *Advances in Reproductive Physiology*. Ed. A. McLaren. London; Logos Press (1967).

Determinants of Infant Behaviour, vols. III and IV. Ed. B. M. Foss. London; Methuen (1967, 1969).

Gonadal hormones and parental behaviour in birds and infrahuman mammals. D. S. Lehrman. In *Sex and Internal Secretions*, 3rd edition, vol. 2. Ed. W. C. Young. Baltimore; Williams & Wilkins (1961).

Animal Behaviour, 2nd edn. R. A. Hinde. London and New York; McGraw-Hill (1970).

Mechanisms of Animal Behaviour. P. R. Marler and W. S. Hamilton. New York; Wiley (1966).

3 Environmental effects
R. M. F. S. Sadleir

The structure of animals and the ways in which their various systems function are the result of selection over millions of years to fit them for the environment in which they live. Like all other physiological processes, reproduction responds to external stimuli so that appropriate numbers of young are produced at the most propitious time of year. This chapter will describe ways in which mammalian reproduction is affected by physical and biological agents.

There are two ways in which environmental factors can govern the most overt expression of reproduction – the breeding season. In 1938, John R. Baker of Oxford first propounded clearly the idea of proximate and ultimate causes:

'Animals have evolved the capacity to respond to certain stimuli by breeding. In cold and temperate climates it is usually clear that the season adopted allows the young to grow up in favourable climatic conditions, and one may say that in a sense these conditions are the *ultimate* cause of the breeding season being at that particular time. There is, of course, no reason to suppose that the particular environmental conditions favourable to the young are necessarily the one or ones which constitute the proximate cause and stimulate the parents to reproduce. Thus abundance of . . . food for the young might be the ultimate, and length of day the proximate cause of a breeding season. Agencies which start and stop reproduction but which do not operate under natural conditions of existence may be called artificial causes.'

Our present knowledge of reproductive ecology more than 30 years later does not alter Baker's basic dichotomy. A number of proximate causes are known and something (although not nearly enough) of the ways in which they effect reproduction, but there is little concrete information on the ultimate causes of breeding in mammals. Although Baker was probably correct in his suggestion that food governs the season of births, almost nothing is known of the relative energetic efficiencies of raising young at one time of the year compared with another.

The proximate causes of breeding are of two types. *Obligatory factors* are natural physical factors which, when changing, *always* cause breeding to start or stop. In most mammals, light is the only such factor, although temperature and other factors can be obligatory for lower vertebrates. *Modifying factors* are natural physical or biotic systems which may influence the response of individual mammals to obligatory factors but which can *only* modify them. This distinction will become clearer in the description of temperature effects below.

LIGHT

In 1924 a Canadian zoologist, William Rowan, working in Edmonton, kept two groups of migratory birds called juncos (*Junco hyemalis*) in outdoor cages during the winter instead of letting them migrate south. In addition he subjected one group to extra periods of artificial light. When birds of this group were killed at intervals until after Christmas, it was found that the testes of males had enlarged considerably and a single female had enlarged follicles in the ovary. This remarkable effect was caused simply by switching a light on and off, despite temperatures that dropped to $-45\ ^\circ$C.

This pioneer experiment initiated two decades of detailed research into the effects of light duration (photoperiod) on the breeding seasons of vertebrates. In 1932 the first papers appeared describing light effects on the reproduction of the field vole (*Microtus agrestis*) and the ferret (*Mustela putorius*) and there is now experimental evidence of photoperiodic control of gonad development in 22 species of mammals representing 8 orders. However, well over half the investigations reported have been on the ferret and the sheep and we still know very little of photoperiodic effects in marsupials, insectivores, bats and primates.

Before describing the experimental evidence for photoperiodic control of breeding, two sets of natural observations are relevant. In a number of mammals the beginning of the breeding season occurs on almost exactly the same date every year. For example, the average date of first ovulation in ewes in Otago, New Zealand, was 15 March, 16 March, 17 March and 21 March over a 4-year study. In species with large ranges the timing of breeding is distinctly related to latitude. Breeding starts later and later in the year and the breeding season shortens as the latitude increases.

The first 'experiments' on photoperiodic effects were not done with any experimentation in mind. It appears that Britons, when colonizing the Southern Hemisphere, were governed by a great urge to take wild and domestic animals from the homeland as fellow colonials. This habit bequeathed to the Antipodes such ecological gems as the rabbit, the fox and red deer. However, the shipping of red deer (*Cervus elaphus*) from U.K. to N.Z. and back produced interesting observations on the alterations in their breeding seasons (Fig. 3-1). Red deer breed on a decreasing light regime so that stags moved to N.Z. bred twice inside a 12-month period. They thus responded to the second period of decreasing light after arrival. Yearling N.Z. females shipped to U.K. came into heat first at the N.Z. time but shifted their seasons to the U.K. pattern in 2 years.

The true experiments into photoperiodic effects have been of four types.

(*a*) Keeping animals for long periods of time in darkness.

(*b*) Keeping animals for long periods of time in continuous light.

(*c*) Gradually altering the light regime so as to keep animals under a regularly decreasing photoperiod when the natural photoperiod is increasing, or vice versa.

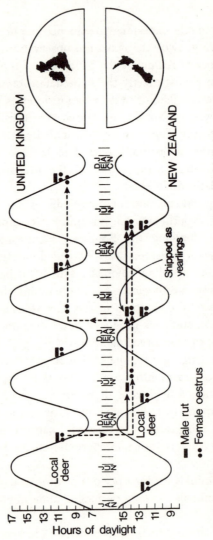

Fig. 3-1. The alteration of breeding seasons of red deer (*Cervus elaphus*) on movement from one hemisphere to another. (Based on data in F. H. A. Marshall. *Biol. Rev.* **17**, 68 (1942).)

(*d*) Placing animals at various times of the year under fixed ratios of light and dark (i.e. 14 hours light:10 hours dark).

The first two types of experiment have produced conflicting results and have really done very little to explain photoperiodic effects on breeding. The third type has demonstrated that a

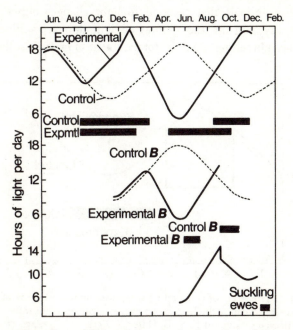

Fig. 3-2. The effect of light regime on the breeding of sheep. The season during which oestrous cycles occurred is indicated by the black band. (Reproduced with permission from N. T. M. Yeates. *J. Agric. Sci.* **39**, 1, text-fig. 5 (1949).)

number of mammals can be made to breed outside their normal season by gradual alteration of photoperiod and nothing else. Fig. 3-2, shows, for example, that sheep will always breed under a decreasing light regime. They can be made to breed twice a year by this method, thus doubling the lamb crop. Most mammals are *long-day breeders* (rodents, carnivores) which start to breed as the days get longer, but some are *short-day breeders* (sheep, deer) whose cycles begin on a decreasing light regime. The fourth class of experiments was originally carried out to determine the effects of constant equatorial light regimes on mammals. They have shown (see Fig. 3-3 for example) that photoperiod does not have to be altered gradually to induce oestrus, and this has lead to considerable controversy as to whether animals that respond to light changes either (*a*) in some way 'add up' the amount of light they are subjected to (and start breeding when the total number of hours is sufficient),

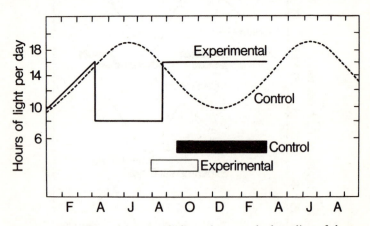

Fig. 3-3. The effect of constant light regimes on the breeding of sheep. The season during which oestrous cycles occurred is indicated by the bars. (Based on data in E. S. E. Hafez.*J. Agric. Sci.* **42,** 232 (1952).)

or (*b*) start breeding when the daylight length 'triggers' some mechanism which required a minimal amount of light in a 24-hour period. The second alternative has been largely discredited by experiments such as that shown in Fig. 3-2. Notice that the two experimental groups of ewes bred in response to different lengths of daylight.

Because the seasonal changes in daylight length are absolutely constant, mammals that respond to photoperiods can start breeding at virtually a fixed date each year. This is probably also true of the cessation of breeding but not many experiments on this aspect have been carried out. Photoperiod is thus a *zeitgeber* (literally, a time giver). If young mammals born at one time of the year survive better than those born at other times, selection will favour those individuals that result from matings at the most appropriate time. If photoperiodic response is genetically determined these individuals will also tend to breed at the most appropriate time.

Two more aspects of photoperiodic effects on reproduction should be mentioned.

(*a*) In sheep and a number of rodents the duration of photo-period can affect the attainment of sexual maturity (puberty). Groups of young voles (*Microtus arvalis*) placed under increasing photoperiods, or fixed long days, reached puberty earlier than groups under decreasing photoperiods or fixed short days. Young female lambs may have their first oestrous cycles at 5 months if this coincides with a period of decreasing light, but in other groups puberty can be delayed by as much as 4 more months or until the first period of decreasing light occurs.

(*b*) In some mammals the blastocyst does not implant until many months after fertilization. This phenomenon is called *delayed implantation* and is described in Book 1, Chapter 4, and Chapter 1 of this book. After the delayed section of pregnancy, implantation in mink (*Mustela vison*) is controlled by photoperiod and this may also be the situation in other species showing this mechanism.

TEMPERATURE

Perhaps the major evolutionary advance of mammals was the development of a regulated internal body temperature – of homoeothermy – which allows them to exploit a wider range of thermal environments than their reptilian cousins, but which leaves them with the problem of maintaining homoeothermy under many different climatic conditions. Generally speaking, reproduction in mammals is only secondarily affected by temperature. Temperature usually has a direct effect on other

75

physiological processes, which in turn alter reproductive patterns.

Extreme cold results in reduced growth rates as most assimilated energy is used in simply maintaining body temperature. Under these conditions puberty in rodents is delayed considerably and oestrous cycles become infrequent or may stop entirely. In nature cold weather can interrupt breeding in species that normally breed continually (e.g. *Rattus norvegicus*) or can shorten the season of seasonal breeders (e.g. cotton rat *Sigmodon hispidus*). If temperatures are warmer than usual in late summer, sheep will start breeding earlier than after a cool summer. Similarly in some rodent populations a cold snap in spring can delay breeding by a few days. Temperature can thus advance or retard the onset or cessation of regular breeding seasons but only within relatively narrow limits. It acts as a modifying factor (see page 70) because eventually breeding will begin (or end) in response to light stimuli.

Very few mammals live naturally in environments where high temperatures present major physiological problems. Research into high temperature effects on reproduction in sheep and cattle reveals that these species must be subjected to extremely high temperatures (or to moderately high temperatures for prolonged periods) before their reproductive processes are inhibited. Under such conditions the male sex drive (libido) decreases and spermatogenesis is impaired. The latter effect is almost certainly due to the physiological difficulties of keeping the scrotal testes a few degrees below body temperature in high ambient temperatures (see Book 1, Chapter 3). Females continue their cycles but prenatal mortality increases considerably, especially before implantation. Such effects in both sexes are found naturally only where domestic stock are reared in high temperature environments. This occurs in parts of Australia and Africa which are thermally very different from those latitudes where the cattle and sheep originated, in Europe or Asia.

A more natural effect of temperature on reproduction is found in hibernating mammals who seasonally become essentially poikilothermic. Hibernation in ground squirrels (*Citellus*) and prairie dogs (*Cynomys*) is regulated to a degree by temperature, but the experimental evidence regarding the way in which temperature alters their breeding is impossible to decipher at the moment. Fortunately, in bats the picture is clearer. The increasing temperatures of spring arouse bats from their winter torpidity and the females then ovulate. Ovulation can be advanced experimentally by bringing bats into warmth in mid-winter.

Finally there is tantalizingly incomplete evidence that human reproduction is affected by temperature. Detailed studies of birth statistics have shown that in temperate regions human conception rates tend to be highest in late spring and summer. Above 21.1 °C or below 4.4 °C conception rates tend to decline. Australian research has shown reduced summer conception rates in tropical areas and an increased summer conception rate in cool temperate regions. Unfortunately nothing is known of the physiological causes of this variation. It must be emphasized that the variation is only small (although statistically significant!) and that there is no area in the world where conceptions do not occur every month of the year.

NUTRITION

The physical factors of light and temperature can alter in very simple ways. Photoperiod can only vary with time while temperature also varies in a simple fashion along a continuous range. Nutrition is a much more difficult variable to measure. An adequate diet consists of sufficient energy intake and sufficient amounts of many different types of protein, minerals and vitamins. Thus, in considering the effects of nutrition on breeding, we are really dealing with qualitative and quantitative variation in at least fifty different dietary components. Often the amount of one constituent (say protein X) will alter or modify the effects of a deficiency of another (say mineral K) so that the potential combinations of effects are truly baffling. To make matters worse we usually do not know what type of reproductive response (see Fig. 3-4) can result from alterations in the total intake or the intake of individual components.

Despite this despairing picture, there are many experimental and field studies which show that alterations in nutrition *do* affect reproduction. The trouble is that we usually cannot determine exactly which nutritional components have caused the result. Perhaps this is why most experimentalists have tried to vary the gross intake by simply decreasing the voluntary food consumption, while keeping constant the relative proportions of the dietary components.

Experimental alterations of the diet

There have been many experiments in which a poor natural diet has been supplemented or the amount of food naturally consumed has been artificially reduced. Such alterations affect a wide variety of reproductive phenomena so that only the major consequences can be detected.

Puberty. Much work has been done on cattle and pigs. Cattle in South Africa fed extra food in winter reached puberty at an average of 440 days compared with 710 days for those not fed extra food. Reduced intake delays puberty. Also cattle, pigs and cats need specific vitamins to achieve puberty. Reduced food intake delays puberty in the rabbit from 7 to 11 months.

Length of breeding season. Reduced gross intake slightly delays onset of breeding in deer (*Odocoileus virginianus*). Green feed is essential for rabbit breeding and the length of breeding season in Australian experiments varied between 77 days (no green feed) and 100 days (pasture only) to 141 days (pasture plus extra green feed).

Proportion of females breeding. The gross quantity of food eaten altered the proportion of females undergoing oestrous cycles in sheep for up to 12 months afterwards.

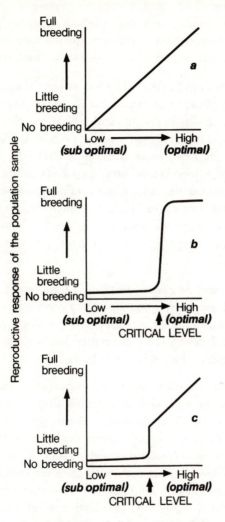

Fig. 3-4. The possible types of reproductive response to nutritional variation in mammals.

Environmental effects

Male breeding. The level of protein intake is particularly important to bulls, rams and probably boars. Vitamin A deficiencies increase abnormalities in spermatozoa and decrease sex drive. Human starvation reduces sperm count and motility.

Litter size. This can be affected by the number of eggs produced and by the number of embryos that die. Starvation can result in prenatal mortality in rodents and rabbits and the number of eggs shed in rats is altered by the protein intake. Sudden increases in the food intake (from suboptimal to optimal) in sheep and pigs just before breeding, a procedure known as 'flushing', causes an increased number of ewes to have twins (133 ovulations/100 ewes on ordinary diets increased to 217 ovulations/100 ewes when flushed). Litter size in rabbits declines steeply (to an average of 2.6/female from 4.8) when no green feed is available.

Birth weight and immediate survival. Low protein intake decreased fawn weight by 2 pounds from 7 pounds in white-tailed deer, and fawn mortality was highest when does were on low protein diets. Ewes on low protein diets had high death rates of lambs, especially when twins were born.

There is thus a wide variety of responses to experimental alterations in the diet. It is unfortunate that so much of this information comes from domestic animals where we know so little about the degree of nutritional depletion they may have endured in the wild (pre-domesticated) state. What little evidence is available suggests that (with the exception of rabbits) most mammals must be subjected to very considerably reduced intakes (perhaps below the minimal natural level?) before reproduction is affected.

Field situations involving artificial supplementation of food

In many areas wild mammals have supplemented their diet by feeding on man's agricultural crops and and this has affected their reproduction. These sorts of observations are particularly interesting because such extra feeding is voluntary and the animals are unrestrained.

The length of the breeding season has been altered by freely selected agricultural supplements in a number of mammals. Muskrats (*Ondatra zibethicus*) can breed later in the year in Iowa when they can feed in nearby fields of corn. A small wallaby, the quokka (*Setonix brachyurus*), is normally a seasonal breeder on an island off the West Australian coast, but females feeding on a municipal dump breed almost all the year around. Similarly, rabbits of the genus *Sylvilagus* breed all year round when feeding on irrigated fields in California and Texas. In Israel a higher proportion of female voles (*Microtus guentheri*) were pregnant and had larger litters on irrigated than on non-irrigated fields. Many more female pocket gophers (*Thomomys bottae*) are found pregnant in Californian alfalfa fields than in other fields. Finally in Manitoba, white-tailed deer which were helping themselves to high protein crops such as alfalfa, proved to have increased numbers of twins, suggesting a flushing effect similar to that seen in sheep.

Comparison of field situations involving natural differences in nutrition

Population studies of wild mammals often show statistical differences in breeding between (*a*) different locations in the same year, (*b*) at the same location but at different times of the year, or (*c*) at the same location at the same time of year but in different years. Often such differences in breeding can be explained in terms of good or bad nutrition – the latter explanation being applied *after* the differences in breeding have been noted. This is bad science unless it is followed by controlled experiments in which natural breeding is altered by supplemental feeding. In the absence of such experiments, we will review briefly the evidence on the way natural fluctuations in food supplies alter reproduction.

Environmental effects

The age of puberty is greatly influenced by nutrition in grazing mammals such as kangaroos and deer. Red kangaroos (*Megaleia rufa*) reach puberty earlier in locations where food is better, and puberty is delayed by as much as 6 months if the animals grow up under conditions of drought. In moose (*Alces alces*), and wapiti (*Cervus canadensis*), variations in nutrition between areas and between years affect the proportion of females that become pregnant when yearlings. Red deer (*C. elaphus*) have reached puberty under very good nutritional conditions in New Zealand at 16 months compared with 3 years on poorer nutrition in Scotland.

Field investigations in Australia, on Antarctic islands, in New Zealand, in the U.S.A. and in Britain and Poland have shown that breeding in various lagomorphs, particularly the European rabbit, is very directly connected to nutrition: in many different environments, they will not breed unless plentiful green feed is available. If such feed is present, breeding may still be inhibited or affected by other factors, but under no circumstances will breeding take place if no green feed is present.

The length of the breeding season can be affected by nutritional availability. For example, red kangaroos start breeding when there is growth of vegetation after rains, and the proportion of non-breeding females slowly increases as drought periods extend. In small forest rodents (*Apodemus sylvaticus*, *A. flavicollis* and *Clethrionomys glareolus*) the availability of acorns, beech mast and other deciduous fruits governs the length of breeding season each year, particularly the time when breeding ends. There is a considerable literature on the different rates of recruitment in white-tailed and mule (*O. hemionus*) deer populations under varying range conditions. For example, mule deer on very poor range produced an average of 0.77 fawns per doe while on very well managed range this was increased to 1.65 fawns per doe. Although there are many reports in tropical climates of the breeding seasons of mammals coinciding with seasonal rains and thus good nutrition, these cannot be considered as evidence for a nutritional effect on breeding. Nutrition in these cases may, however, be the ultimate cause of seasonality in breeding (see beginning of chapter).

Reproductive energetics

There is a whole branch of ecology concerned with the patterns and efficiency of energy flow through ecosystems. Each individual female mammal absorbs energy in her diet and either uses such energy to maintain her basic metabolism, stores it in her own tissues or transfers it to her offspring. Unfortunately we

TABLE 3-1. Energy requirements of two rodents under different conditions

	Bank vole (*Clethrionomys glareolus*)	Common vole (*Microtus arvalis*)
Rates of assimilation (kcal/day)		
Non-reproducing female	17.5	11
At the end of pregnancy	24.0	16
At the end of lactation	40.0	38
Total energy required above maintenance (kcal)		
For pregnancy	75	60
For lactation	290	243

know very little indeed about the ways in which such energy is used in reproduction although consumption rates give an indication of the extra demands of pregnancy and lactation. Ewes that are not lactating consume from 800 to 1100 grams of digestible organic matter per day, while lactating ewes consume from 1500 to 2500 grams depending on the type of food utilized.

Work in Poland and Czechoslovakia on two species of wild rodents indicates increased energy requirements as shown in Table 3-1. These figures are based only on food assimilated and do not include the amount of energy derived from initial fat reserves. There is a very great need for more investigations of this type in other mammals. Apparently 11 per cent of the extra energy absorbed by bank voles during pregnancy is transferred into the protoplasm of newborn young, and $14\frac{1}{2}$ per cent of the extra energy taken in during lactation is also transferred into the flesh of the suckling babies. For common voles the figures are 14 and $15\frac{1}{2}$ per cent.

Nutrition and human reproduction

Unfortunately for the understanding of reproductive processes, no one experiments on human beings. It is even more unfortunate that there are so few records of human reproduction under conditions of malnutrition or subnutrition. The future pattern of population change of our species may well depend on our breeding responses to nutritional deficiencies. There are a few reports that extreme malnutrition results in a decline in the birth rate but nothing seems to be known of the stage at which reproduction is affected. Human beings continue to breed at amazingly high rates under conditions of subnutrition and even malnutrition. What little information is available suggests that the human response differs from that of other mammals, where reproduction is generally reduced progressively as nutritive intake decreases. In terms of Fig. 3-4 most mammals seem to respond as in graphs *A* and *C* whereas humans respond as in *B* with a very low critical level. A high reproductive rate despite poor nutrition is an unfortunate species characteristic.

There is so little information on these matters that what has been suggested above is speculative, but we have more data on nutritional effects during lactation in our species. Research in Africa and other parts of the world has shown that lactation alone does not reduce a woman's fertility. If she is suffering from poor nutritive intake, her lactation is still maintained (perhaps for even longer periods) but she does not resume menstrual cycles for a prolonged period and thus is infertile. Thus lactation under conditions of malnutrition in man can reduce fertility but there is a need for much more data on this and other effects of nutrition on human reproduction.

When many more investigations on nutrition, and especially on energetics, have been carried out we will begin to understand why some species breed at different times of the year and why some produce many young and others only one. Nutrition is generally supposed to act as the ultimate factor in regulating breeding in mammals. We should not be surprised therefore that, for most mammals, nutritive intake in the wild is optimal for reproduction, because the breeding season has resulted from selective pressures that caused such seasonality. Although the season of breeding may be genetically fixed, nutrition can have a variety of effects on the production of young in different areas and can thus play a major role in the recruitment of young into populations.

SOCIAL FACTORS

Mammals develop more complex social structures than any other form of animal. With few exceptions, they go through life having repeated contact with other members of their own species. In order to exploit their communal environment to maximum advantage, they develop special relationships with each other (for example hierarchies) which allow their societies to function smoothly. There is a wide range of interreactions and this is reflected in the many types of society that can be seen. Compare the pattern and movements of a cohesive band of baboons (*Papio*) always under the sway of a few dominant males, with the looser association of a herd of red deer occasionally responding to a dominant hind. The organized society of a prairie dog (*Cynomys*) town, where each animal roams freely inside his band's home range (but is in real trouble if he ventures into that of another band), is quite different to the freedom with which bands of gorillas (*Gorilla*) mix when they come into each other's home range. Individual social contact amongst adult mammals extends from the extreme unsociality of the wolverine

(*Gulo gulo*) and the kangaroo rat (*Dipodomys merriami*) which only come together for a very short time at oestrus, to the highly social house mouse (*Mus musculus*) or rhesus monkey (*Macaca mulatta*) where it is almost impossible for individuals to get away from one another.

Social contact is thus an essential part of the external environment of a mammal and such contacts may completely dominate daily activities and even responses to other environmental factors. Alteration in the social environment results in dramatic changes in behaviour patterns and sometimes these are reflected in physiological changes. Individuals of species which are naturally fairly social usually have a greater tolerance to such changes than individuals of more solitary species. Because of species differences, we will consider separately the role of social factors in reproduction for the three groups of mammals that have been most investigated.

Mice

Ease of manipulation, rapid breeding, and their ability to be kept in large numbers in relatively small areas, have made the house mouse the 'most investigated' species as far as social effects on reproduction are concerned. Because of their association with human agricultural activity, mice are often found in great numbers in barns, food stores and ricks. At high densities, increased contact between individuals means increased social pressure. Dense mouse populations in ricks have a much-reduced reproductive potential. Young mammals show inhibited puberty in dense populations and the average litter size declines significantly.

Two types of experimental situations have demonstrated other effects of density on mice breeding. In the first, mice have been kept in small cages in various density and sex combinations, and their individual reproductive patterns followed closely; while in the second, large populations have been allowed to grow until densities were reached such that reproduction was inhibited in some way. To deal with the first situation – when small groups of females (four to five) are kept in small cages, there is an increased incidence of pseudo-pregnancy (see Book 1 Chapter 4) and oestrus may be delayed for over a week. This is called the *Lee–Boot effect* after its discoverers. In even larger groups (ten to thirty) oestrous cycles become irregular and many females stop cycling altogether. However, if males are introduced into cages containing numerous females the average length of the oestrous cycles returns to normal. Another fascinating effect is that the females appear so delighted with the males' presence that many of them start coming into oestrus in synchrony. This is called the *Whitten effect* and is most obvious if the female density was particularly high prior to the introduction of a lucky male. Similar synchronization of oestrus can also be shown in sheep, goats and cattle. A final social effect on mouse breeding does not require conditions of overcrowding. If the male to which a female was mated is removed as soon as she becomes pregnant, and a strange male is introduced into her cage within the next 4 days, she gets very upset, and the blastocysts inside her do not attach to the wall of the uterus. Her pregnancy is thus blocked and she comes back into oestrus – a phenomenon called the *Bruce effect*.

We are now fairly certain that the causative agent in the above effects is the smell of the male urine. This urine contains *pheromones* – substances secreted by individuals which can result in specific physiological reactions in other individuals of the same species. In female mice this male odour may stimulate pituitary production of FSH and LH so that follicular development occurs, sometimes earlier than otherwise expected; it may block pregnancy by inhibiting pituitary prolactin release and so causing the corpora lutea to regress.

If groups of mice are allowed to breed unrestrictedly in population pens with plentiful food and shelter, the numbers inevitably rise, but eventually an upper limit is reached. At this time recruitment of young into the population stops. Often, very few young survive until weaning, owing to abnormalities in maternal behaviour or to disturbance by too many neighbours. In addition, the birth rate of the population as a whole declines. The most important factors contributing to this decline are inhibited puberty at high densities, increased prenatal mortality (i.e. between conception and birth) and a reduced number of cyclic females in the population.

With the possible exception of the Bruce effect, all of these reproductive changes in mice are seen at very high densities. Most of the laboratory population work has been done at densities far higher than are ever found in natural mouse populations in houses and barns. These commensal populations in turn, living in areas of plentiful food, are much denser than wild or feral mice populations. Most wild rodents are never found at such densities, and we may infer that in nature social or other mechanisms have acted to keep densities below the level at which reproductive processes will be affected.

Rabbits

Although there have been many studies on mouse behaviour we know little in this species of the role of an individual's social position in determining his (or her) reproduction. Fortunately extensive Australian studies of penned rabbits have correlated density and its effects on individual sociality with reproductive success. Hours of painstaking observations of floodlit pens, where the behaviour of each individual rabbit has been studied in detail, has given considerable insight into the importance of social hierarchies and territorial behaviour in rabbit reproduction.

The earliest work showed that as populations were allowed to increase in pens the normally continuous breeding slowed down. Instead of a high proportion of does conceiving a litter at the oestrus immediately following birth, which is the normal process, some does did not breed for many days after giving birth and, in most cases, as the density increased, this gap in breeding became longer and longer. There was also an increased rate of resorption of embryos in the uterus. Workers could not agree that the dominant does had less resorption or a less interrupted breeding pattern than submissive does, but there was no doubt of the importance of hierarchial position in regulating the production of kits. Dominant does obtained the best territories in the pen, tended to have larger litters, and dug

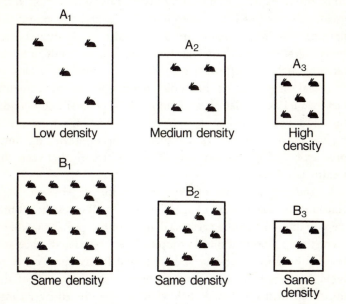

Fig. 3-5. Varieties of arrangement of rabbit densities in pen experiments. (Based on data in K. Myers, C. S. Hale, R. Mykyowycz and R. L. Hughes. In *Behaviour and Environment*. Plenum Press, New York (1971).)

their breeding stops in advantageous positions, so that all their kits survived. Submissive does were pushed out to harder ground or ground more subject to flooding. Their litters were smaller and had poorer chances of survival.

Recent investigations by Myers and his colleagues in Australia have been into the role of space (with and without density changes) in rabbit breeding. The reproduction and behaviour of rabbits in the six types of pens shown in Fig. 3-5 have been studied, there being a number of replicates of each arrangement. In the A series the absolute numbers were kept constant but space and density altered. Rabbits normally live in a small social unit of 1-3 males and 1-5 females which defend a common territory, and in this experiment such units were maintained in all of the A pens. A significantly higher rate of ovulation and a greater number of litters were born in A_1, and the amount of fighting was much greater in A_3. Females in A_3 dug shorter breeding stops and their kits died mainly as a result of social factors, such as disturbance by strange females or a lack of maternal care, while kits in A_1 died mainly because of ecological factors such as predation or flooding. The general body condition and body weights of embryos and nestlings were much higher in A_1 than in A_3. Despite all these effects, the total production of young did not differ significantly between the treatments.

In the second series of experiments the density was kept constant but the amount of space varied. In all cases the rabbits were maintained at the greatest density of the previous experiment. In B_3 a normal social group had very restricted space and the frequencies of aggressive and sexual behaviour were very high. In B_1 several social groups were formed, but all of the measured forms of female behaviour were reduced over those prevailing in the other two pens. Again the total productivity of young did not differ between treatments. The length of breeding burrow was shortest in B_1 and most of the kits there died as a result of social disturbances while those in B_3 died because of ecological factors.

Perhaps the most significant results were those showing the overall relationship between individual social status and breeding in all of the females used in these experiments. Dominant females survived better and had higher body weights. More importantly they produced significantly more litters and more kits than submissive females. Other results showed that young rabbits born to mothers who were crowded were poorer in physical condition and tended to behave differently from young born to uncrowded mothers.

This fascinating research has demonstrated that reproductive responses of rabbits to crowding are not due solely to increased density, but also to some unknown features of space – perhaps the ability of individual rabbits to organize themselves socially in space. There can be little doubt that similar mechanisms exist in other social mammals. Their reproductive responses to density changes are really only responses to alteration in their normal relationships to other individuals, owing to increased densities.

'Cyclic' mammals

A characteristic feature of the population numbers of microtine rodents (the vole family) is the occurrence of cycles of abundance and scarcity. In any one place numbers will be relatively low for 1–3 years, but 1 year in 4 these numbers will increase enormously, perhaps 50-fold, only to crash the subsequent year. This is known in lemmings (*Lemmus lemmus, L. trimucronatus, Dicrostonyx groenlandicus*) and voles (*Microtus agrestis, M. californicus, M. montanus, M. pennsylvanicus, Clethrionomys glareolus, C. rufocanus, C. rutilus*). Detailed investigations of reproduction under naturally occurring differences in density are perhaps the best indication of the ways in which social factors may alter breeding. Unfortunately we know very little of the types of behavioural alterations that occur during these cycles. There appears to be increased aggression in the peak and decline years but how this behaviour is distributed amongst the individuals in the population is unknown.

During peak years the attainment of puberty is much delayed and some individuals born in one season will not reach puberty inside that season – a very peculiar phenomenon for small mammals like rodents. The length of the breeding period also alters during the cycle. Although breeding intensity may decline in the autumn of the year when numbers are increasing, considerable numbers of females produce young under the snow all through the winter, so that in the spring of the peak year reproduction is already going full blast. And then in the peak year itself breeding usually stops much earlier than normal. There is little alteration throughout the population cycle in the size of litters produced, or in the proportion of adult females breeding.

A few mammals (muskrat, *Ondatra zibethicus;* snowshoe hare, *Lepus americanus* and possibly black-tailed jackrabbit, *L. californicus*) show much longer, 10- or 11-year, cycles of abundance. The length of this cycle has made it very difficult to study but there is good evidence that the litter size in these mammals does alter with the phase of population change. In muskrats litter sizes are larger in high density years and smaller in low density years.

Mammals in general

Before concluding this review of social factors in mammalian reproduction, one more phenomenon deserves mention. In a number of highly social species, such as certain large primates, seals, and deer, the adult males are extremely aggressive towards younger males that have just reached sexual maturity. In particular they drive them away from prime oestrous females so that these young males do not actually breed until a number of years later than they could potentially mate. For example, young bull elephant seals (*Mirounga leonina*) are physiologically capable of mating at 4 years of age but often do not do so until 6 to 15 years old.

Ethology is a relatively new and greatly expanding branch of biology and more and more is being learned (particularly from detailed and time-consuming field studies) of the complexities of mammalian social organizations. Copulation itself is a social phenomenon, and during lactation the young mammal is continually learning his relationship with adults of his species. Disturbance of social patterns can thus easily lead to reproductive malfunction. Our own species is exceptional in that much of its behaviour is culturally adapted, but it would be rash to assume that man has no unlearned behaviour. Such innate behaviour must have been selected for during the 99 per cent of the history of our species when we lived as small semi-nomadic family bands of hunters and food gatherers. We really know little of the extent of such behaviour, and thus there is little we can predict of the way in which man will behave reproductively under the extreme conditions of high density and social stress that seem now to be threatening the race.

SUGGESTED FURTHER READING

External factors in sexual periodicity. E. C. Amoroso and F. H. A. Marshall. In *Marshall's Physiology of Reproduction*, vol. 1, pt 2. Ed. A. S. Parkes. London; Longmans Green (1960).

Evolution of breeding seasons. J. R. Baker. In *Evolution – Essays on Aspects of Evolutionary Biology*. Ed. G. R. De Beer. Oxford; Clarendon Press (1938).

Vertebrate Reproductive Cycles. W. S. Bullough. London; Methuen (1963).

Reproduction and Environment. R. L. Holmes. London and Edinburgh; Oliver & Boyd (1968).

Bioenergetics of pregnancy and lactation in the bank vole. F. Kaczmarski. *Acta Theriologica* **14,** 409 (1966).

Bioenergetics of pregnancy and lactation in the European common vole. P. Migula. *Acta Theriologica* **14,** 167 (1969).

The effects of varying density and space on sociality and health in mammals, with special reference to the wild rabbit. K. Myers, C. S. Hale, R. Mykyowycz and R. L. Hughes. In *Behaviour and Environment*. New York; Plenum Press (1971).

Comparative Biology of Reproduction in Mammals. Ed. I. W. Rowlands. Symposium No. 15 of the Zoological Society of London (1966).

Biology of reproduction in mammals. Ed. I. W. Rowlands and J. S. Perry. *Journal of Reproduction and Fertility* Suppl. 6 (1969).

The Ecology of Reproduction in Wild and Domestic Mammals. R. M. F. S. Sadleir. London; Methuen (1969).

4 Immunological influences
R. G. Edwards

Various defences exist in the body to combat infectious or other diseases, and an important form of protection is afforded by our immunological responses to foreign substances. The responses occur rapidly, and the defence is powerful and specific. Immunological responses are involved in a number of different ways with various aspects of reproduction and fertility. In this chapter we shall be particularly concerned with these matters, but first we should look briefly at general features of the immune system.

There are two main types of immunological response. One involves the formation of antibodies which pass into the blood stream or other body fluids, and which are known as circulating antibodies. The second type involves the participation of cells such as lymphocytes, which also respond to immunization. Antibodies have been shown to discriminate between molecules that are closely similar in structure, or to react selectively with small amounts of a particular chemical in a mixture. Compounds that react with antibodies are called antigens. The chemical nature of antigens is very varied – some are simple compounds, others are complex proteins or polysaccharides forming part of the membrane of a virus or a cell. There may be several antigens in a single virus for example, or even in a single molecule. The wide diversity in the cells of the body results in a widespread assortment of antigens, some being common to many tissues while others are restricted to only one type of tissue. Lymphocytes react with one or more specific antigens after being 'instructed' to do so under the influence of macrophages, the scavenging cells that scour the tissues for matter foreign to the body. Lymphocytes are also involved in reactions such as the rejection of transplanted organs or tissues. The breadth and scope of immunology has changed in the last twenty-five years. Its earlier successes were in studies where the unique nature of antigens carried on viruses or other agents made them susceptible to attack without risk to the person being immunized. We are now aware of the delicate and balanced immune responses ever-present in ourselves, and how the situation must be disturbed as gently as possible. Immune reactions against ovarian tissue, for example, can ultimately destroy all the Graafian follicles. While respecting the power and discriminating nature of antibodies, we must also pay great heed to the progressive immune responses that can arise naturally or following experimental treatments.

The label 'antigen' describes a wide variety of chemicals that exists either in solution or as an integral part of living organisms such as viruses, bacteria and the cells and tissues of higher animals. Some antigens are now chemically defined; for example, the structure of the insulin molecule has been described in detail and clearly recognizable parts of it are known to induce the formation of different antibodies. Most antigens, however, have not been identified chemically, but their reactions with antibodies are so well characterized that various inferences can be drawn about their molecular size and structure as well as their similarities with other molecules.

THE CHEMICAL NATURE AND OTHER PROPERTIES OF ANTIBODIES

Antibodies belong to the group of proteins known as globulins. They are found in blood and secretions, and on mucous surfaces throughout the body. A great deal has been learnt about antibody molecules in recent years. Some antibodies will cause bacteria or other cells to agglutinate into large clumps, or to burst, and these are known as agglutinins and lysins, respectively. Others, known as precipitins, cause soluble antigens to become visible as a precipitate, while yet others, the antitoxins, will react with toxic chemicals and make them harmless. The molecules of many, though not all, antibodies have two active sites, both directed against the same antigen. This structure confers on antibodies the power to cause agglutination because they are able to combine with two antigenic molecules simultaneously. Some of the reactions causing lysis are complex and may involve other components of the blood which complement the action of the antibodies; indeed, the word 'complement' has become accepted as a simple term to describe these agents.

The chemical nature of antibodies was elucidated in the 1960s. There are now known to be five classes of antibody molecules in man, with a basically similar chemical structure (Fig. 4-1). Each antibody is composed of two pairs of 'chains', each chain being a long string of amino acids. The longer pair are known as the 'heavy' chains and the shorter as the 'light' chains. All four chains are held together by chemical bonds involving sulphur atoms. The antibody molecule can be broken at a particular point by digestion, with an enzyme such as papain, to yield fragments identified as F_{ab} and F_c fragments. The F_{ab} fragments possess the antibody activity (the 'antigen-binding fragments'), and the F_c fragments have various other properties (the 'crystallizable' fragment). The ability to fragment antibodies in this way has led to many fundamental studies on the properties of the different parts of the molecules.

Chemical and physiological differences distinguish the five classes of human antibodies. The differences largely reside in the chemical composition of the heavy chains, and in the manner in which separate molecules join together. The 'natural' ABO antibodies of the blood-group system are examples of one class – these are larger molecules than the others. When a person is immunized, a smaller antibody appears in the blood; for example, if a group A man is immunized with red cells carrying antigen B, he produces 'immune' anti-B in addition to the natural anti-B antibody he had before. A third class of antibody appears to act in defence of surfaces such as the lining of the mouth, the bronchial ducts and other areas that are covered by mucous membrane. The fourth class has the property of becoming fixed to the surface of tissues, and takes part in reactions such as allergies (e.g. hay fever, or the spasmic response, anaphylaxis). The fifth class of antibody has been recognized chemically, but its role is at present unknown. For simplicity these antibodies,

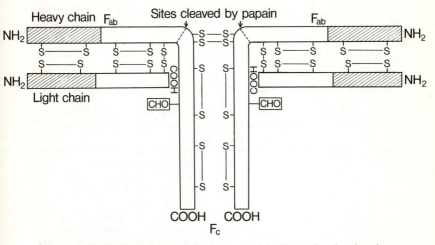

Fig. 4-1. Typical structure of the immunoglobulin molecule, showing the association between the two heavy and two light chains. Papain splits the molecule into the F_{ab} and F_c pieces shown. The N- and C-terminal ends of the chains are indicated by NH₂ and COOH, respectively; in the shaded area at the N-terminal ends the composition of amino acids varies. These areas are believed to carry the specific activity of each antibody towards a particular antigen. The unshaded areas are constant in their amino-acid composition and sequences for each species; these parts are believed to play a role in the transport of antibodies across tissues, the fixation of complement, attracting antibody to cell membranes, etc. (From G. M. Edelman. 'The structure and function of antibodies'.) *Sci. Amer.* **223,** 34, top fig. on p. 38. © 1970 by Scientific American, Inc. All rights reserved.)

or immunoglobulins as they are called, are given the prefix Ig, and the class is designated by a third letter. The five classes just described are shown in Table 4-1 and are labelled IgM, IgG, IgA, IgE and IgD respectively.

The five immunoglobulins are of different sizes, and this can affect their properties. Thus IgM has a high molecular weight

97

because it consists of five molecules joined together, offering ten binding sites. The large size of IgM is probably responsible for its failure to be transmitted from a mother to her fetus. In IgA, two molecules are joined together. The chemical nature of these antibodies confers other characteristics on each of them, and some of their effects will be described during the following

TABLE 4-1. The classes of immunoglobulins and their role in the body

Immunoglobulin class	Molecular size of the antibody in body secretions	Type of immunological activity associated with each class
IgM	19S	Agglutinating, involved in 'natural' antibodies against blood-group A, etc. Fails to cross placenta
IgG	7S	Produced in response to immunization. Traverses placenta and other membranes
IgA	11S	Found at mucous surfaces in addition to serum; it may defend these surfaces from infection or other antigens
IgE	–	Attaches to tissues and is responsible for allergies or anaphylaxis
IgD	–	Role as yet unclear

discussion. Differences in molecular size and other physical properties enable us to separate antibodies and determine their relative quantities in biological fluids. (Figs 4-2 and 4-3.)

Various antibodies have been found to react with tissues of the reproductive tract. Some of these antibodies arise naturally, while others can be induced experimentally in man and animals by immunization. The pathological significance of many of these antibodies is obscure; some seem to be innocuous, others evidently have disastrous effects on the animal or on the fetus. The effects of particular antibodies cannot be predicted. Some antigens are well protected behind cellular or basement mem-

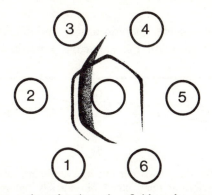

Fig. 4-2. The number of antigens in a fluid can be revealed by allowing it to diffuse towards a specific antibody, and then identifying the number and density of the resulting precipitation 'lines'. A supporting medium, e.g. agar, is used to permit the antigens and antibodies to diffuse together from separate wells. In this example, antiserum taken from rabbits immunized with dog prostatic fluid was placed in the centre well. Various fluids taken from different tissues under analysis were placed in the outer wells: 1 dog prostatic fluid, 2 dog prostate gland, 3 dog bladder, 4 dog kidney, 5 dog serum, 6 saline. One thin precipitation line formed continuously between wells 1–5 and the centre well, showing that one antigen was present in, and common to, each of these respective tissues. A second broad line formed between wells 1–3 and the centre well, showing that a different antigen was common to these three tissues only, and absent from the others. The breadth of this line, especially against well 2, indicated that there may be several antigens in it, and further analysis showed this to be the case. (From C. Yantorno, S. Shulman, M. J. Gonder, W. A. Soanes and E. Witebsky. *J. Immunol.* **96**, 1035, fig. 2 left. © 1966 The Williams & Wilkins Company, Baltimore.)

branes, and so escape combination with their corresponding antibody. Some antibodies are produced in the central immunological tissues of the body, and are distributed throughout the body, whereas others are produced or act at limited and widely dispersed sites. Antibodies may exert their action only at certain times in the reproductive cycle, either because the levels of antibody vary or because of the transient appearance of the antigen. Generalizations about the effects of antibodies are thus difficult to make, and each example has to be examined separately.

We will now consider the appearance and effects of various antibodies in relation to the particular reproductive process

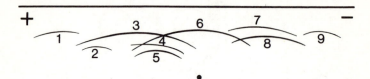

Fig. 4-3. Greater discrimination between different antigens can be obtained by separating them electrophoretically before adding the antibody. This method is known as immunoelectrophoresis. A minute drop of the fluid to be analysed is placed in agar dissolved in a buffer of known pH. This point is marked by the dot. An electric current is then passed, which separates the antigens along a linear pathway according to their electric charge. Antibody is then added to the trough (shown by the continuous line), and the antibody and antigens diffuse together. The number of precipitation lines reveals the number of antigens, and their positions show how far the antigens had moved during electrophoresis. Information on the number of antigens and some of their physical properties is thus obtained. In the example shown above, human seminal plasma was analysed with an antibody against it that had been prepared in rabbits. Nine precipitation lines were found. (From A. Hekman and P. Rumke. *Fert. Steril.* **20,** 312 fig. 1g. © 1969 Hoeber Medical Division, Harper & Row.)

involved, dealing first with antibodies in the male and then with antibodies in the female reproductive tract. Later, immune reactions during pregnancy will be described, especially in the 'acceptance' of a fetus by its mother. This leads naturally to a fourth section on the transmission of protective antibodies from mother to young, followed by a penultimate section on the rhesus factor. Finally, we will briefly consider some experimental work with antibodies against hormones and other molecules.

REPRODUCTION IN MALES

Both naturally occurring and induced antibodies against their own testicular tissue have been found in male animals and in men. Broadly speaking, there are two major consequences in immunized males. First, testicular cells and immature spermatozoa in the testis might be destroyed, a phenomenon generally called autoimmune aspermatogenesis. Secondly, spermatozoa might be agglutinated or lysed after ejaculation so that the fertility of the male is suppressed. The two responses thus differ with respect to the types of cells that are damaged, and perhaps also in the antigens involved and the nature of the reactive antibody.

Autoimmune aspermatogenesis

This is an example of the destructive nature of some antibodies. The condition arises as a result of experimental immunization. When males are immunized with pieces of their own testis or with testicular tissue from another member of their species, the resulting immunological reaction is so violent that many of their germinal cells are destroyed. Spermatids, spermatocytes and spermatozoa are lysed, and even spermatogonia too if immunization is sufficiently heavy (Fig. 4-4). Guinea pigs are among the most sensitive animals, although similar damage can be induced in other species, including man. The destruction might be localized in particular tubules or restricted regions of the testis, but it can become so widespread that large areas are depleted of spermatogenic cells. Fertility is obviously impaired under these conditions, although spermatogenesis is resumed after several weeks or months, and fertility is then restored. At one time there were high hopes that a male contraceptive might be developed by inducing this reaction in man, but optimism was premature. Destroying large areas of testis over a long period is of dubious merit, and the side effects of the treatment – including abscess formation at the site of immunization – are hardly a

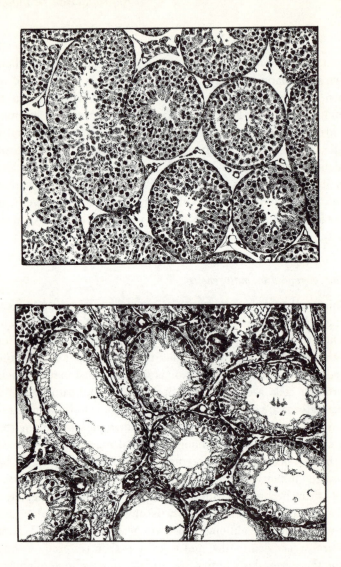

Fig. 4-4. Destruction of germinal cells in guinea pigs immunized by testis material. A control testis is shown at the top, and all stages of spermatogenesis are present. The effects of immunization are shown at the bottom: many of the tubules are depleted of spermatozoa, spermatids and spermatocytes. (From R. G. Edwards. *Sci. J.* April, 69, by courtesy of D. B. A. Symons. Two figs. at top of p. 71 (1967).)

recommendation for acceptable contraceptive procedure. Moreover, we do not fully understand the causes of the lesions in the testis, and more knowledge is needed before we can contemplate its adoption for man. Libido is not impaired, indicating that the endocrine system is not unduly disturbed; consistently, Leydig cells generally escape the wave of destruction.

We are uncertain about the nature of the testicular response to immunization. Some sceptics maintain that the antigens need not come from the testis, and that various other tissues can be substituted. Many studies have shown the testis to be uniquely sensitive to disruption by diverse external agents and that spermatogenesis is very prone to interruption. Nevertheless, there is strong evidence pointing to some sort of specific immunological response being involved. Several antigens have been so purified that testicular lesions are caused when only a few micrograms of any one of them are used for immunization. These antigens are proteins or glycoproteins, most of them deriving from the acrosome of the spermatozoon. The availability of such powerful antigens is a great stimulus to research.

It has also proved difficult to decide which immunological processes are involved in autoimmune aspermatogenesis. Two or three years ago, most scientists believed that damage was induced by immunologically active cells rather than by circulating antibodies, for damage could be provoked in a normal male by transferring cells from an immunized donor (Fig. 4-5). Moreover, testicular damage appeared to be correlated with other types of immune response known to be mediated by cells. Sensitized lymphocytes were believed to expose the germinal cells to the destructive action of antibodies and complement.

But doubts have been expressed about this interpretation. Questions first arose about the route whereby the antibodies, circulating and cellular, reached the germinal cells. We now believe that they gain access indirectly, via the rete testis, rather than directly across the basement membranes of the tubules, for leucocytes and other cells accumulate around and in the rete testis soon after immunization. Direct transmission into the tubules is largely blocked by the efficient basement membrane surrounding the tubules and the 'tight' junctions between Sertoli cells, both structures preventing the direct access of antibodies and cells until after the initial damage has taken

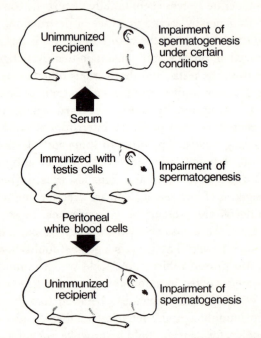

Fig. 4-5. At one time, immunologically active cells were believed to be the major cause of autoimmune aspermatogenesis. This figure shows that cells transferred from a donor guinea pig to a second animal can impair spermatogenesis (lower part of figure). Recently, however, serum from immunized guinea pigs has also been shown to induce lesions under certain circumstances. (From R. G. Edwards. *Sci. J.* April, 69. Fig. at lower right on p. 71 (1967).)

place. The role of cellular responses has also been called into question because circulating antibody alone was shown to invoke testicular lesions. Damage to the testis of normal guinea pigs could be induced by injecting antibody against spermatozoa into the rete testis. It is also possible under some circumstances to induce testicular damage by transfusing normal guinea pigs with serum taken from a previously immunized animal (Fig. 4-5).

Some other observations also cast doubt on the idea that sensitized cells alone induce testicular damage. A natural antibody has been found in the serum of normal animals of many species: it will lyse and destroy spermatozoa and germinal cells *in vitro*. Why such an antibody exists in non-immunized males (and in females – even virgins) is difficult to explain. This antibody might cause damage *in vivo* if it gained access to the tubules. It has been identified in guinea pigs as belonging to a class of antibody known to participate in other destructive reactions. There is obviously a great deal to learn before the mechanisms involved in autoimmune aspermatogenesis are fully understood.

Infertility in men due to autoantibodies against spermatozoa

A different immunological response, so far found mostly in man, is the suppression of fertility by the agglutination or immobilization of spermatozoa after ejaculation (Fig. 4-6). The agglutinated spermatozoa cannot ascend the female tract. The condition arises naturally but is fortunately rare – some 3 per cent of infertile men possess these autoantibodies. The condition probably arises through temporary or partial blockage in the vas deferens, or inflammation in the male tract, and once established it is difficult to reverse. The antibodies can be detected both in blood and seminal plasma. Some of these men can evidently become temporarily fertile when they are given hormones to suppress spermatogenesis; the antibody titres decline as the production of spermatozoa ceases, and then, when the hormone treatment is stopped, spermatozoa reappear before the levels of antibody rise again.

Circulating autoantibodies against spermatozoa are also found in many men after vasectomy, i.e. the surgical removal of part of the vas deferens. Spermatogenesis continues after the operation, but as the spermatozoa can no longer travel down the vas deferens they are resorbed instead. These conditions evidently lead to autoimmunization of men by their own spermatozoa. Circulating antibodies do not inevitably impair fertility in these men, for repair of the vas deferens has enabled some of them to have their own children.

Fig. 4-6. Spermatozoa are agglutinated or immobilized by some anti-bodies. In this illustration, mouse spermatozoa have formed a network of agglutinated cells after serum containing antibodies was added to them.

There has been a great deal of debate as to how antibodies suppress male fertility. If circulating antibodies are not responsible, as shown by the restored fertility after repair of the vas deferens, those in seminal plasma could be the cause of sperm agglutination. It is thus necessary to know how they enter the male reproductive tract. There seem to be two possibilities: they might actually be synthesized in the tract if immunization in fact occurred there, or if immunization occurred systemically, the antibodies could be transmitted into the reproductive tract. Antibody levels in seminal plasma and in serum vary in different men, although more is usually found in serum. Antibodies are evidently associated in some way with the prostate gland, for they tend to be localized in the prostatic fraction of the ejaculate. The antibodies in serum and in semen actually belong to different classes of immunoglobulins. IgA sperm-agglutinins seem to be present in higher titres in seminal plasma, whereas more IgG spermagglutinins are found in serum. IgA is typically found at mucous surfaces (Table 4-1), and is either synthesized or effective at local sites. The infertile men might therefore have two sources of antibodies against spermatozoa, systemic immunization having led to circulating IgG, and local immunization to seminal IgA.

Similar findings on circulating antibodies against spermatozoa are now being made in vasectomized animals, but there are no examples yet of local effects in the accessory glands of male animals.

The known infertility of men with antibodies again raised hopes of the development of an immunological contraceptive. Perhaps one day this approach will be successful, but there are many difficulties to be overcome. How is the prostate (if this is the gland concerned) to be stimulated to produce local antibodies? Nor are the long-term effects of these and other sperm-agglutinating antibodies well documented, and at present it would obviously be wise to be cautious about this approach to the control of fertility.

Analyses of seminal plasma and spermatozoa

We must ask various questions about the relationships between the antigens present in spermatozoa and seminal plasma, and the immune system of the female. Are blood-group antigens, or those responsible for inducing the rejection of grafts, the transplantation antigens, carried on spermatozoa? Can the presence of these or other antigens in semen lead to the immunization of the wife after repeated intercourse? These and other problems have provoked a good deal of study into the antigenicity of semen. Let us consider spermatozoa and seminal plasma separately.

Many studies on spermatozoa have been concerned with the blood-group and transplantation antigens. The observations that have been made contribute to the solution of another problem: are these antigens determined by the haploid genotype of the spermatozoon or by the diploid phenotype of the spermatogenic cells? Results obtained so far have been somewhat conflicting. First, claims that the ABO antigens exist on spermatozoa have been vigorously denied. An aspect that was insufficiently appreciated in many studies was the effect of the 'secretor' gene. Men and women secretors produce soluble ABO antigens in their body secretions, such as saliva (while non-secretors do not), and the secreted antigens then stick on to cells. Male secretors have AB substances in their seminal plasma, so that spermatozoa are coated and give 'false' positives. This kind of result undoubtedly misled some workers into the belief that the ABO antigens were actually present in the membrane of the spermatozoon. In other experiments, very weak positive results were noted with spermatozoa from non-secretors. Current belief is that these antigens are either not expressed on human spermatozoa or are carried in very small amounts.

Transplantation antigens might also be present on spermatozoa or in seminal plasma, as shown in studies with human and mouse. Moreover, the haploid genotype of individual spermatozoa could exert some effect over the expression of these antigens. Immunological methods might therefore be useful in identifying and separating individual types of spermatozoa carrying particular antigens. Some people believe that repeated mating of female animals can result in the female becoming sensitized against transplantation antigens on spermatozoa. But it is so easy to be misled in this field – at one time infertility due to ABO incompatibility between husband and wife was believed to arise through immunization by spermatozoa, and this view is no longer widely held.

Seminal plasma has also been examined immunologically. Attention has been concentrated on the large number of diverse antigens in this fluid. Some of the antigens in it are also found in other body tissues; for example, albumin and some globulins are found both in serum and semen. Other antigens are found only in seminal plasma, and their source can be traced to a particular gland (Figs. 4-2 and 4-3). Six antigens in the seminal plasma of guinea pigs arise in the seminal vesicle, and some of these are also found in prostatic fluid. Few of the antigens in seminal plasma have yet been characterized chemically, although one found in human semen is known to be an iron-binding protein immunologically identical to lactoferrin in milk. This antigen adheres tightly to spermatozoa, and dominates their antigenic behaviour. Several of these 'sperm-coating antigens' exist, like the blood-group substances in human secretors described above. Antibodies against coating antigens affect the gamete less adversely than do antibodies directed against the spermatozoon itself: immobilization of spermatozoa, for example, is not caused by antibodies against seminal plasma, and male fertility is unimpaired. In contrast, antibodies against spermatozoa may be lytic or inhibit motility, and can suppress fertilization under some conditions.

REPRODUCTION IN FEMALES

The immunological system of the female is challenged in various ways during intercourse and pregnancy. Spermatozoa, seminal plasma and the fetus all introduce foreign antigens into the mother, and the fetus enters into prolonged close contact with maternal blood after implantation. Reactions to the fetus could therefore differ qualitatively and quantitatively from those occurring against semen and will be described separately. First, however, we must consider an example of autoimmunity in which women react against their own tissues and destroy their follicles and oocytes.

Immunological influences

Autoimmunity to ovaries and adrenals

Many patients with an illness known as Addison's disease and one or two with Cushing's syndrome produce autoantibodies against antigens of their own adrenal glands. The reaction can then spread to other tissues, such as the thyroid, perhaps because antigens similar to those in adrenals also exist there. Sera obtained from 5–10 per cent of women with Addison's disease contain autoantibodies that will react with steroid-producing tissues in ovary, placenta, and even testis. Antibodies against the ovary are probably of most significance clinically, for the patients may become amenorrhoeic and infertile through the autoimmune destruction of their follicles.

Several different antibodies appear to react with ovarian tissue. Initial responses seem to involve circulating antibodies. Later, lymphocytes and other cells invade and destroy the follicles. Large follicles are the most sensitive, perhaps because they contain most antigen, but ultimately even the smallest are destroyed so that no follicles are left. This example of the disastrous consequences of an immunological reaction serves as a warning against the too-easy adoption of techniques of contraception depending on antibodies.

Antibodies in females against spermatozoa and seminal plasma

Antibodies against spermatozoa and seminal plasma in the female tract have been found less often than might be expected, even where continuous and repeated immunization may occur, as in prostitutes. Some evidence indicates that the seminal antigens are degraded enzymatically, so averting an immune response. Alternatively, the vast number of leucocytes that invade the uterine lumen in the post-ovulatory phase and phagocytose the spermatozoa might confer the necessary protection. The invasion also occurs during a cycle without coitus and in virgin females, so is not attributable only to the presence of spermatozoa.

Nevertheless, naturally induced antibodies against semen have been found in women. One woman suffered from an extreme response to a seminal antigen, for she experienced a violent uterine anaphylaxis during intercourse with her husband. The anaphylaxis became generalized, with the risk of death. The antigen was identified as a glycoprotein in the semen; the antibody was presumably an IgE, the immunoglobulin mediating similar types of allergic sensitivity elsewhere in the body. Uterine antibodies against spermatozoa have not been detected in women, although the problems of collecting uterine fluid make these investigations somewhat difficult. In animals, extensive search has revealed uterine antibodies against bacteria, but not unequivocally against spermatozoa. Antibodies against seminal antigens and spermatozoa were found in the uterine cervical secretions of some women, and were considered to be IgG. IgA could also be involved in the cervix since it is known to protect mucous surfaces. Follicular fluid contains immuno-globulins that are probably derived from the circulation, for this fluid is closely in equilibrium with blood plasma. Antibodies in follicular fluid could thus enter the oviduct at ovulation, and might exert their action in the female tract or confer some immunological protection to it.

Many women are reputedly infertile because they possess circulating antibodies against spermatozoa, but many of these claims appear to be ill-founded. Circulating antibodies against spermatozoa and seminal plasma can be induced experimentally in female animals and in women. Fertility in animals is often impaired after immunization, although there is doubt about which physiological system may be influenced by the antibodies. A positive correlation is found to exist between infertility and the titre of circulating antibodies in some experiments, but not in others. Fewer spermatozoa enter the oviduct after immunization and the resulting decline in the rates of fertilization appears to be sufficient to account for the lowered fertility. The major problem has been to detect the effect of antibodies on spermatozoa in the uterus. The relative importance of immobilization, agglutination or phagocytosis of spermatozoa is not understood. There is some evidence to suggest that prostitutes develop tissue-fixed antibody against spermatozoa in their uteri, which might lower their fertility. These antibodies are likely to be produced locally rather than systemically. Fertility in animals is impaired by antibodies specific for spermatozoa rather than by those directed against antigens in seminal plasma. There is some evidence to suggest that antigens in semen can maintain an immune response in females given an initial immunization. Regular mating of immunized animals maintains the levels of circulating antibodies against spermatozoa, and conversely the use of a condom is reported to reduce the titres of these antibodies in women.

There is thus some possibility for fertility control by inducing in women antibodies that react with spermatozoa. Difficulties with this method, as judged from animal experiments, will be the variation in titre between women, the sporadic effect of these antibodies on fertility, and our current failure to understand the basic mechanisms affected by the antibodies. Clinical application must be considered distant.

Antibodies in pregnancy

Eggs and embryos have many antigenic components. Natural antibodies occur in some species against eggs; such antibodies are indeed found in some women with Addison's disease and Cushing's syndrome as mentioned earlier. Fertilization can be suppressed by the use of antibodies. Blastocysts and the trophoblast on older embryos and fetuses also have antigens that are susceptible to various antibodies.

The aspect of pregnancy that has perhaps stimulated most interest is the survival of a fetus in an antigenically distinct mother (discussed also in Book 2, Chapters 1 and 3). Why is the fetus not rejected as a foreign graft? The first point to be decided is whether conditions exist for rejection to occur. One necessary condition is that the transplantation antigens be present on the surface of trophoblast and other tissues exposed to maternal fluids. There is some doubt that this is so, for the methods used to detect these antigens have been indirect, e.g. by studying the development of embryos transferred to ectopic sites. Conclusions drawn from this type of work may not be relevant to the situation existing in the uterus. In addition, direct immunological tests on the trophoblast of early embryos have yielded contradictory data.

Perhaps the decidual response in the tissues of the uterus itself modifies the immune response of the mother, and protects the fetus. If so, then similar responses must presumably occur in ectopic sites such as the testis, brain and kidney, which are known to sustain embryonic and fetal development to advanced stages of differentiation. Indeed, the testis and kidney will accept an implanting embryo more readily than the uterus itself, where nidation will occur only at a limited time after ovulation. There are some indications from embryo-transfer experiments that an immune response is indeed involved in implantation. Rat embryos usually fail to implant in the mouse uterus, but they can be induced to implant and develop to mid-term if tolerance to rat tissues is first induced in the female mouse.

There seems no doubt that the uterus can respond immunologically to foreign antigens. Grafts placed there are rejected, although this is not necessarily a reliable guide to the situation in pregnancy since the anatomical relationships that arise between graft and uterus are very different from those between mother and fetus. On the other hand, a mother will reject a cutaneous graft of skin while accepting fetuses of a similar genotype in her uterus. Many explanations have been offered to account for the protection of the fetus, and some of them are now known to be dubious, e.g. those postulating that steroids in pregnancy induce a generalized decrease of the immunological responses in the mother.

Many immunologists believe that there are definite interactions between mother and fetus. Others maintain that the fetus is protected simply by the exclusion of the mother's lymphocytes, and possibly her complement, by the placental barrier. There are, in general, two schools of thought among those who believe that interactions exist. One school postulates that 'low-dose tolerance' of the fetus arises in the mother. This reaction has been studied mostly in relation to skin grafts: the graft is not rejected as long as the host is constantly given small numbers of cells genetically like the graft. When treatment stops, the graft is rejected. The continual passage of small numbers of fetal cells or antigen might suppress rejection by the mother in a similar way, so protecting the fetus.

The second school of thought believes that a different immunological response occurs in the mother, namely 'enhancement facilitation'. This is a paradoxical reaction that has been studied in other areas of research. Two forms of response occur upon immunization: a cellular reaction which is rejective, and a humoral reaction that actually confers protection against rejection. Enhancement is believed to be involved in various other phenomena, such as in the growth of some cancers. The principles underlying enhancement are now being applied to organ transfer in man as an alternative to invoking tolerance or suppressing rejection by antilymphocytic serum. The mechanisms involved are not clear. The circulating antibody might coat the foreign antigens and so prevent the lymphocytes from recognizing them as foreign, or the circulating antibody might reduce the degree of the cellular response to the antigens. Enhancement, rather than rejection, might also be a response to a particular distribution or concentration of antigens on the surface of a graft.

What is the evidence that each or any of the mechanisms exists during pregnancy? Lymphocytes taken from a pregnant mother are evidently reactive against fetal cells, yet circulating antibodies collected from her at the same time will protect the fetal cells. Protection by serum antibodies is even strong enough to resist the onslaught of lymphocytes taken from other animals that had been highly immunized against fetal antigens. This dual response of the mother is typical of enhancement facilitation, and argues against the induction of low-dose tolerance. Other data lend indirect support to theories involving enhancement facilitation, for protective antibodies induced in the mother can evidently be transferred into the fetus.

Immune interactions between mother and fetus can be shown to occur in other ways. Paternal skin grafts or cancer cells, which are normally rejected, will apparently be accepted by the mother towards the end of pregnancy. Repeated pregnancies can induce the permanent acceptance of male tissue by female mice, provided the antigenic differences between the two parents are minor. The mother's sensitivity is reduced mostly by the presence of fetuses, for only slight effects have been ascribed to the antigens in semen. In contrast, the mother can actually become sensitized to fetal tissue under some conditions: for example, a rhesus-negative woman can become immunized against her rhesus-positive fetus (we will return to this topic later). Mothers can also be immunized against transplantation antigens carried by the fetus: sera used for typing human transplantation antigens are often obtained from multiparous women.

Although enhancement or low-dose tolerance may play a role in maintaining pregnancy, some doubt remains about just how relevant these responses are. Enhancement requires that both circulating antibodies and sensitized cells in the immunized recipient are in direct contact with the antigens of the graft. Normally few maternal lymphocytes colonize the fetus, although some authors have reported extensive colonization. The dearth of immunologically active maternal cells in the fetus would therefore exclude enhancement facilitation, in the accepted sense, from conferring any protection. At best, only those membranes exposed to the maternal circulation would need protecting. We have in effect returned to theories that exclude any role for protective antibodies, relying instead on various physiological barriers between maternal and fetal tissues. A layer of fibrinoid material has been described as such a barrier covering the transplantation antigens of the trophoblast in human and mouse placentae. Physiological barriers, and the absence or weak expression of transplantation antigens on trophoblast, might thus be sufficient to confer protection during a normal pregnancy. We need to know more about the expression of antigens on the fetus and its membranes, and about maternal responses, before definite conclusions about protective mechanisms can be drawn.

The transmission of immunity from mother to young

At birth, the neonate has a weakly developed immunological system. Its capacity to make antibodies is limited. It has lived in the mother's uterus, sheltered from many infections and other foreign antigens by maternal antibodies. Only under certain circumstances, such as the occurrence of uterine infections during pregnancy, can the fetus be shown to produce antibodies, and even then the response is poor and restricted largely to IgM antibodies. The defence that the neonate thus needs for a few days or weeks after birth, while its own immunological systems develop, is conferred by maternal antibodies transmitted to it during or immediately after pregnancy. Hence the array of antibodies developed by the mother are conferred on the neonate, so providing a vital defence against infection until they deteriorate after some days or weeks.

Differences exist between species in the means used to transmit maternal antibodies to the fetus (Table 4-2). In some species, maternal antibody is transferred entirely before birth, the route being via the yolk-sac placenta or the allanto-chorionic placenta. In others, maternal antibody is absorbed entirely via the colostrum, the milk first produced by the mother. Others again utilize both methods, antibody being transmitted during and after pregnancy. We will consider three species, rabbit, cow and man, to illustrate the principles involved.

Transmission in the rabbit occurs during pregnancy, and has perhaps been most widely studied. During early pregnancy, antibodies enter the yolk-sac cavity of the embryo and thence pass to the vitelline circulation across the membrane known as the bilaminar omphalopleur; this membrane breaks down later in pregnancy, and transmission is then through the yolk-sac splanchnopleur. Maximum levels of maternal antibody in fetal blood are reached shortly before birth. All types of antibody are transmitted, of both large and small molecular weight, although there is some selectivity in the rate of transfer. This can best be shown by injecting antibodies from other species into the mother, and studying how quickly they reach the fetus. Under these circumstances, rabbit antibodies are transferred

TABLE 4-2. Transfer of immunity from mother to offspring

Animal	Stage of development when transfer occurs	Route by which transfer is established
Rabbit	All prenatal	Yolk sac
Cow	All postnatal	Colostrum via gut of newborn
Man	Almost all prenatal	Placenta
Rat	Some prenatal, most postnatal	Yolk sac prenatally, milk postnatally via gut of offspring

more rapidly than, for example, human antibodies. The fragments of antibodies obtained by digesting immunoglobulins with enzymes are also transmitted: the F_c fragments enter the fetus rapidly, whereas F_{ab} fragments enter slowly if at all. This information suggests that the F_c fragment is responsible for ferrying the whole molecule across the fetal membrane.

The situation in calves is quite different: transmission of antibodies occurs through the first milk or colostrum, and not before birth. The rate of transmission is astonishing, for the antibodies must be transferred within a few hours of birth.

Immunological influences

This is necessary because certain enzymes soon develop in the gut of the calf and these can digest the immunoglobulins. The colostrum is so rich in antibodies – far richer indeed than the mother's blood – that a few hours suffice for the accumulation of high levels of antibody in the calf's blood. As in rabbits, there is some selectivity in the types of antibody transmitted, and IgG is found in colostrum in highest concentration. Most of the immunoglobulins are concentrated into colostrum from the mother's blood, and are not synthesized in the udder. After suckling, all of the immunoglobulins in colostrum are absorbed with equal facility from the alimentary canal of the calf; the control of selective transmission thus resides in the secretion of antibody into colostrum and not in the selectivity of the neo-natal gut. The level of maternal antibodies in the calf declines to half the original by approximately 40 days of age, but traces are still present for up to 9 months or longer. It is worth noting that in mice and rats, where transmission occurs both before and after birth, transfer via milk continues for days or weeks, until the gut becomes impermeable to antibodies.

Antibodies are transferred to the human fetus in yet another manner. Transmission occurs overwhelmingly during pregnancy, but the route is unlike that in the rabbit. This is not surprising since the membranes surrounding the human fetus are of a different origin and structure than are those around the rabbit fetus. Most transmission in man occurs via the allanto-chorionic placenta, and numerous studies have been made on blood taken from the umbilical cord at birth. Many antibodies, although not all of them, are found in cord blood at a concentration similar to that in maternal serum. Transmission evidently increases towards the end of pregnancy, for small infants reputedly possess less antibody than larger babies.

As in the other species, there is some selectivity in the classes of antibodies that are transferred. An excellent example is provided by the transmission of the blood-group antibodies. Natural ABO antibodies are IgM, whereas 'immune' ABO antibodies are IgG. Examination of infants has shown that immune antibodies are transmitted readily, but the natural antibodies are not transferred at all. Transfer of immune antibodies can lead to difficulties for those fetuses possessing different ABO or rhesus blood groups to the mother, as we will see later in this chapter. The levels of antibodies against viruses such as poliomyelitis can actually be higher in the fetus than in the mother. On the other hand, antibodies against bacteria living in the alimentary canal are weakly transmitted to the fetus. Indeed, the transfer of maternal antibodies conferring defence to the infant gut might be a rare example of transmission via colostrum in man, for the IgA in human colostrum might act as an immediate defence for the child's alimentary canal when suckling begins. Some antibodies might also be transmitted in low amounts via a third route, i.e. across the fetal membranes into the amniotic fluid, and then swallowed by the fetus. The human fetus can synthesize some of its own antibodies, and levels of IgM and IgA in the fetal circulation might indicate some degree of synthesis by the fetus. Levels of IgG are largely decided by maternal transfer. The levels of maternal antibody in the infant's blood decline to approximately one-half of the original amount by 25–30 days after birth and sufficient amounts of antibody are usually synthesized by the baby from some 3 months or so after birth.

Similarities could exist between immunological mechanisms involved in the transmission of antibodies, anaphylaxis, and the degradation of immunoglobulins. One theory suggests that the F_c fragment of the antibody is attached to receptors on the surface of cells, and the whole antibody with its receptors is then transported across the embryonic membrane in a vesicle (see Fig. 4-7). Antibody not attached to the receptors is degraded. The degree of transmission is thus restricted to the number of available receptor sites. The F_c fragment is also involved in attaching antibodies to cell surfaces during anaphylaxis, and also carries that part of the molecular structure involved in the degradation of the antibody. The cellular system might thus be similar in all three phenomena.

Immunological influences

Immunization against the rhesus antigen during pregnancy

We have seen how the mother can become immunized against fetal antigens, and also how maternal antibodies are transmitted to the fetus. Should the mother become immunized against her own fetus, she can then transmit antibodies that seriously impair the normal development of her child. The best known example concerns the rhesus (Rh) factor. Some 85 per cent of people

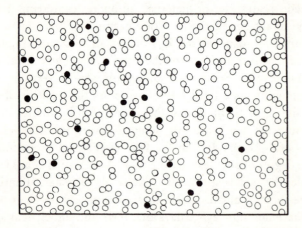

Fig. 4-7. The presence of fetal cells in the mother's circulating blood is a clear warning that a placental haemorrhage has occurred and that rhesus immunization could ensue. The fetal cells can be detected by a simple test for a particular type of haemoglobin, which reveals them as dark dots among the mother's clear cells. Fetal cells enter the mother mostly during labour. (From D. R. S. Kirby. *Adv. Reprod. Physiol.* **3,** 33, plate 6 (1968).)

carry the rhesus factor on their red blood cells. If a fetus with this antigen (Rh +) implants in a mother lacking it (Rh −), the mother can produce and transmit antibodies that can cause irreversible damage to the fetus. The condition is known as erythroblastosis or haemolytic disease of the newborn. It occurs also with other blood-group factors in addition to rhesus. ·

The rhesus factor is not composed of a single antigen: there are several, although one, called Rh(D) is more important than the others in the context of haemolytic disease. Pregnancy is fortunately not a sufficiently strong stimulus for immunization in most women, and only some 5 per cent of them develop antibodies. Usually, the first child escapes the consequences of immunization, but not the second or later children. The off-spring can suffer in various ways, ranging from minor defects in babies to intrauterine or prenatal death. Fetal death occurs in more than 10 per cent of women previously sensitized against rhesus, and liveborn fetuses may develop jaundice and suffer permanent brain damage. The effects of rhesus immunization of the mother have been averted by performing a blood trans-fusion on the fetus *in utero*, or on the child immediately after birth.

Haemolytic disease also arises with ABO incompatibility between mother and fetus, but only when the mother possesses IgG antibodies as a result of immunization in a previous preg-nancy. The naturally occurring IgM antibody does not cross the placenta, and is harmless. Immunization of the mother occurs where the fetus is a secretor belonging to group B, or a certain type of group A, and the mother is group O. The condition affects only a few of the fetuses (1 in 3000 births, or only one-tenth of those suffering from rhesus diseases) and is rarely severe. An important difference exists between the expression of the ABO and rhesus systems, the Rh antigens being fully developed in the fetal tissues whereas AB antigens are only weakly developed. Diseases similar to haemolytic disease also occur in relation to antigens carried in other cells of the blood, e.g. platelets, but these conditions are very rare.

Immunization of the mother leading to haemolytic disease probably occurs by the leakage of fetal red cells across the placenta. Leakage of large numbers of cells can occur during parturition, but some can also enter the maternal circulation earlier in pregnancy (Fig. 4-8). Maternal antibody is often found in the amniotic fluid of the fetus.

Measures are being developed to prevent haemolytic disease of the newborn. In the last few years, immunization of the

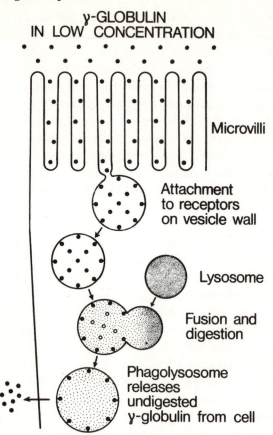

Fig. 4-8. The transmission of γ-globulins might occur via receptors on the cell surface which are carried into the cell in a vesicle. The vesicle fuses with a lysosome, and any antibody not attached to the receptor is digested. If the concentration of antibody is small, as shown on this page, most of it is transmitted; if the concentration is high, as shown on p. 123, most of it is digested. The amount transmitted in both examples remains constant. (From F. W. R. Brambell. *The Transmission of Passive Immunity from Mother to Young*. North-Holland, Amsterdam. fig. 11.1 (1970).)

mother has been prevented by methods designed to stop the fetal red cells from reaching the mother's antibody-forming sites. These developments stemmed from various observations,

γ-GLOBULIN
IN HIGH CONCENTRATION

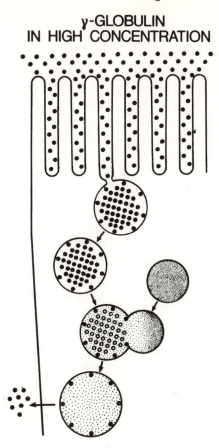

Fig. 4–8 (*continued*)

one of them being the partial alleviation of Rh haemolytic disease if the mother and fetus were simultaneously incompatible for ABO differences; under these circumstances only 10 per cent of the mothers were immunized instead of 17 per cent. The ABO differences probably helped to remove fetal cells more rapidly from the mother's circulation, so reducing the chances of her immunization. An alternative explanation is that the ABO-incompatibility results in the fetal cells being expelled via the liver, a poor producer of antibodies, rather than via the spleen which is highly productive.

Rhesus-negative mothers carrying Rh + babies are now given anti-Rh antibody obtained from women previously immunized against their fetus, or from male Rh− volunteers who have been immunized with Rh + cells. It was first given to mothers at risk, and especially to those with large numbers of fetal cells in their circulation. Experience has shown that the antibody can be safely administered during pregnancy or after the mother has given birth and will prevent her immunization. This evidence indicates that most transplacental haemorrhage responsible for immunizing the mother in the first place usually occurs during birth, and would explain in turn why the firstborn children are very rarely affected. The importance of this discovery is revealed by statistics showing that haemolytic disease of the newborn was formerly the most common form of birth anomaly. It is necessary to be ready in advance and be aware which mothers are likely to have Rh + babies. Highly active antibody must be used, and evidently combines with the Rh antigen on fetal cells, so suppressing immunization of the mother. By separating the IgG fraction of the immune serum, reaction in the mother to other serum proteins has been avoided without reducing the protection conferred by the antibody. Some questions of technique remain to be resolved: will full protection be conferred should large leakage of fetal cells occur earlier in pregnancy, and will sufficient quantities of protective antibody be available from male volunteers when there are few women donors who have been sensitized by pregnancy?

ANTIBODIES AS EXPERIMENTAL TOOLS

Antibodies are so discriminating and powerful that they are often invaluable for work in various scientific and clinical disciplines. They have recently proved highly useful in endocrinology and are being widely used to measure (assay) levels of hormones in blood and other tissues. The techniques for assay are rapid and sensitive, and can be automated. A brief description of this fast expanding field of work is highly relevant to a discussion on antibodies in reproductive processes.

Antibody methods for assaying blood levels of pituitary hormones have mostly been designed for luteinizing hormone (LH), which is responsible for inducing ovulation in the middle of the oestrous or menstrual cycle. Antibodies can easily be raised against proteins like LH, and purified so that they are specific for the hormone. It is then a relatively simple matter to measure to a fine degree the amount of hormone in blood, or in other solutions, that reacts with a standard amount of the antibody. This type of analysis has shown that a large surge of LH appears in the middle of the normal cycle, and lasts for 2 or 3 days. Levels before and after the surge are low. Measurements have also been made on women taking contraceptive pills. Some types of pill were found to completely suppress the LH surge, and so prevent ovulation from taking place. Other pills induced a different response, for repeated surges of LH were detected in the women at intervals of 3 or 4 days, which obviously raises questions about the mode of action.

Steroid hormones can be assayed by similar methods. These hormones are of much smaller molecular size and are therefore weakly antigenic. The difficulty in obtaining antibodies is overcome by conjugating the steroid with a protein, and then using the complex for immunization. Antibodies are produced against the steroid and the protein, but the anti-protein antibody can be removed from the mixture so leaving the antibody specific for the steroid. Assays for steroids have also proved very sensitive – amounts of oestradiol as low as 5×10^{-12} g (i.e. 5 picograms) can be measured – and have permitted various studies to be made; for example, the oestrogen 'surge' responsible for stimulating implantation has now been precisely defined in rats.

Antibodies against hormones have also been used in experimental work, such as for the inactivation of hormones in blood or in tissues. The action of LH, for example, can be suppressed by antibody, so that ovulation or implantation is prevented. Ovulation in rats can also be prevented by using antibodies against oestradiol, and this evidence supports the idea that oestrogen is necessary for triggering the LH release in mid-cycle. Perhaps one day antibodies such as these will be used for contraception, but at present they interfere with too many physiological systems to be of value for this purpose. Moreover, if they are used too frequently the recipient might produce her own antibodies against them and reduce their effectiveness. Examples of such a counter-response have been found in animals, and show that antibodies must be used cautiously if their action is not to be impaired.

Immunology, unlike endocrinology, has developed largely isolated from studies on reproduction, and knowledge relevant to reproduction has arisen in distant biological disciplines. The recent discovery of antibodies responsible for allergic sensitization and for the protection of mucous surfaces, for example, has had immediate significance for studies on conception and implantation. Sometimes new knowledge questions the basis of explanations that were previously accepted. Even while this chapter was being written, the widely-accepted principle of 'tolerance' came into question, because the unresponsiveness of mice to antigens previously given to them at birth seemed explicable as enhancement rather than tolerance. The study of pregnancy has also contributed to our understanding of enhancement – an example of a way in which studies on reproduction help with the analysis of fundamental immunological principles. This interrelationship between immunology and reproduction is bound to come into sharper focus, given the current pace of expansion of knowledge in these two disciplines.

SUGGESTED FURTHER READING

Immunological aspects of pregnancy. D. R. S. Kirby. In *Advances in Reproductive Physiology*, vol. 3. Ed. A. McLaren. London; Logos Press (1968).

Suggested further reading

The foetus as an allograft: the role of maternal unresponsiveness to paternally-derived foetal antigens. C. A. Currie. In *Foetal Autonomy*. Ciba Foundation Symposium. London; Churchill (1969).

Immunology of conception and pregnancy. R. G. Edwards. *British Medical Bulletin* **26**, no. 1, 72–8 (1970).

Immunology and Reproduction. Ed. R. G. Edwards. International Planned Parenthood Federation (1969).

Immunology and Development. Ed. M. Adinolfi and J. Humphrey. London; Spastics International Medical Publications, in association with Heinemann Medical (1969).

The Transmission of Passive Immunity from Mother to Young. F. W. R. Brambell. Amsterdam; North-Holland (1970).

5   Ageing and reproduction

C. E. Adams

Gerontology, the study of ageing, came into being as recently as 1950, according to Alex Comfort in his article on 'The causes of ageing'. And so it is hardly surprising that there are still many gaps in our knowledge of ageing, both in a general context and specifically in relation to reproduction. Apart from what we know of man, satisfactory data on the relation between lifespan and reproduction are available for only a few short-lived laboratory animals and for certain wild animals kept in zoos. Records of farm animals tend to be misleading as only very select individuals are usually retained for breeding purposes, the majority being killed for a variety of reasons. Even with no selection, the decline in a breeding population tends to be very rapid, as shown in Fig. 5-1. Investigation of

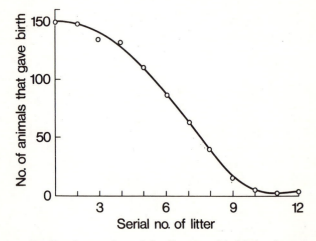

Fig. 5-1. Decline in number of fertile rats with birth of successive litters. (From D. L. Ingram, A. Mandl and S. Zuckerman. *J. Endocr.* **17**, 280, fig. 1 (1958).)

the mechanisms of ageing is plagued by the expense of keeping experimental animals for long periods during which many inevitably die without contributing to the results.

THE COURSE OF SENESCENCE

In litter-bearing animals, such as the mouse, litter size is related to litter order and maternal age; there is an initial rise, a plateau

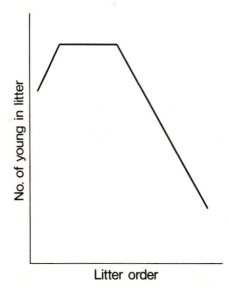

Fig. 5-2. Diagram showing the general relation between litter size and litter order in individual female mice. The reproductive life history consists of an initial rise, a plateau period and a period of linear decline. (From J. D. Biggers, C. A. Finn and A. McLaren. *J. Reprod. Fert.* **3**, 313, text-fig. 3 (1962).)

period and subsequent decline. This is illustrated diagrammatically in Fig. 5-2. In practice, the decline is rarely abrupt, but occupies a significant part of the animal's reproductive lifespan; actual examples appear in Fig. 5-3. During the decline phase, a decreasing litter size is merely one manifestation of a general decay in reproductive function; other signs include changes in the frequency and regularity of oestrous or menstrual cycles and, in monotocous species particularly, a decline in the chances of becoming pregnant.

There are several possible explanations for reproductive decline. Helen King in 1916 attributed it to a reduction in the number of eggs shed, and also questioned whether 'abnormal ova are more frequent in old animals than in young ones and so help diminish fertility in later life'. Yet another

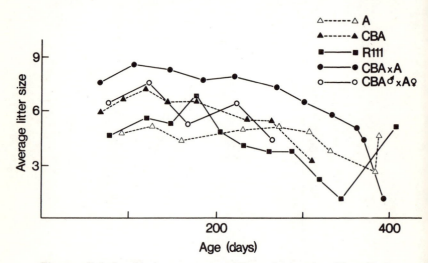

Fig. 5-3. Relationship between age and litter size in mice. (From E. C. Jones and P. L. Krohn. *J. Endocr.* **21**, 469, text-fig. 12 (1961).)

possibility was put forward by Asdell and his co-workers of Cornell, namely, that the accessory organs may wear out so that eggs cannot be fertilized or embryos carried to term. More recently, Esther Jones and Peter Krohn in Birmingham established, from painstaking studies on oocyte populations in the mouse ovary, that the loss of fertility was not due to exhaustion of the oocyte store. In man, on the other hand, oocytes are nearly totally depleted at the time of the menopause. Jones and Krohn concluded that the reason for decline in fertility in the mouse was 'more likely to be found in the hormonal control of the ovary or in the uterine environment'.

Still more recent work indicates that the decline is due to increasing embryonic mortality rather than to a fall-off in ovulation rate, and we now believe that the prime fault lies in the ageing uterus. Experiments on mice, hamsters, and rabbits have proved beyond doubt that the uterus of aged mothers cannot support the development of eggs derived from young donors. In two of these species, rabbit and hamster, it appears that eggs obtained from old mothers are also suspect for very

TABLE 5-1. Survival of fertilized eggs to birth after transfer between animals of different ages. (From C. E. Adams. *J. Reprod. Fert.* Suppl. **12**, 1–16 (1970).)

	Rabbit	Hamster	Mouse
	Eggs surviving to term (%)		
Young to old	1.5	8.3	14
Young to young	50.0	49.2	48
Old to young	12.5	4.5	54

few were able to develop in the healthy uteri of young mothers (see Table 5-1). The somewhat discordant note struck by the apparent success of the mouse eggs, 54 per cent of which survived to term, may simply mean that the donors were relatively younger.

Discussion of the ovarian contribution has been renewed following the report of Alan Henderson and Robert Edwards of Cambridge on chiasma frequency, which led to their 'production-line' hypothesis. They propose that eggs are ovulated in the order they are formed, and that later formation during oogenesis carries greater risk of chromosomal abnormality.

This leaves the abnormal forms to be ovulated in the ageing animal when the oocyte store is becoming depleted.

In ageing rabbits fertilization is less reliable than in young animals where it is common to find 100 per cent of the eggs fertilized. Additionally, in eggs penetrated by spermatozoa a two- to seven-fold increase in cleavage failure has been observed (see Fig. 5-4). We may well ask, to what extent are these expressions of inherent defects in the egg or embryo attributable to ageing changes in the maternal environment? By transferring unfertilized eggs obtained from old animals shortly after ovulation to young recipients it has been found that the primary defect lies within the egg.

In the female mammal no single component of the genital system appears able to escape the consequences of senescence, with certain organs or functions being affected earlier or more severely than others. Species and individual variation undoubtedly exist. Until very recently one could only speculate on their relative significance, but now for the first time it is becoming possible to draw up a balance sheet, albeit one that later may have to be amended.

Owing to fundamental differences in gametogenesis and steroid hormone synthesis the effects of ageing on reproduction in the male are far less obvious than in the female. Unlike oogenesis, spermatogenesis continues from puberty throughout life, and both the 'quality' of the ejaculate and fertility appear to be quite well maintained. In man especially there are several well authenticated cases of the aged begetting offspring. Recent work on male mice suggests that their infertility – most were sterile by 24 months of age – may have been due to physical incapacity rather than to reduction in fertilizing power of spermatozoa. In the bull there is evidence that fertility falls with age, even among highly fertile animals retained for AI; here the physical aspect can be excluded. Over 60 000 artificial inseminations during 1946–9 showed greatest fertility at 3–4 years of age and then a slow decline of about 1 per cent per year (Fig. 5-5).

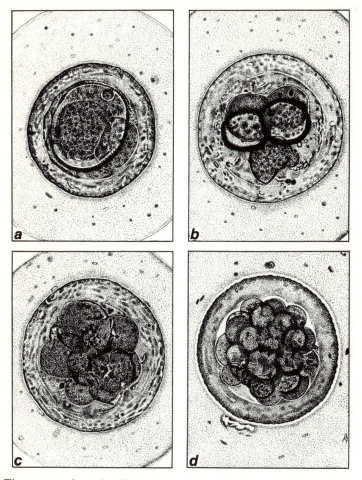

Fig. 5-4. *a*, *b*, and *c* Examples of eggs showing arrested cleavage recovered from the oviduct of an old rabbit 60 h *post coitum*.
d A normal egg (morula stage) recovered from the oviduct of a young rabbit 60 h *post coitum*. (From C. E. Adams. *J. Reprod. Fert. Suppl.* **12,** 1, plate 1 (1970).)

GAMETES

There is a growing awareness of the potentially harmful effects of gamete ageing, particularly of the human egg. Ageing may occur at several different stages during the interval from oogenesis to ovulation, or after ovulation, if sperm penetration is delayed.

With few exceptions the process of oogenesis in mammals is already complete at the time of, or very soon after, birth and the stock of oocytes will not be added to later in life. Consequently, in women an oocyte may have been present in the ovary for 40 years or more before it is ovulated. One of the

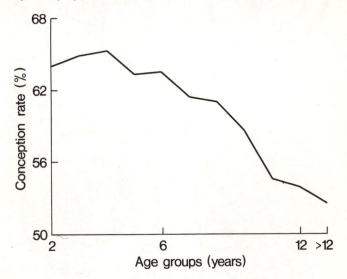

Fig. 5-5. Fertility of bulls in relation to age. Fertility is measured as conception rate at 3 months. (From M. W. H. Bishop. *J. Reprod. Fert. Suppl.* **12,** 65, text-fig. 3 (1970).)

best known and most tragic conditions that could be attributable to ageing is mongolism, or Down's syndrome, due to trisomy of chromosome 21. The incidence of mongolism is of the order of 1 in 2000–2500 for mothers under 30 years, 1 in 1200 between 30 and 34 years and 1 in 300 at 35–39 years, with a progressively steeper rise after this age. The pattern appears to be the same irrespective of racial, geographic or socio-economic factors. The incidence of mongolism in 834 Japanese mothers as related to age is shown in Fig. 5-6. The exact cause of this age-dependent increase is not known. Before the 'production line' hypothesis, the most favoured explanation was that ageing during the prolonged dictyate stage might progressively impair the proper placement and attachment of chromosomes on the spindle during the first meiotic division. Another possibility is

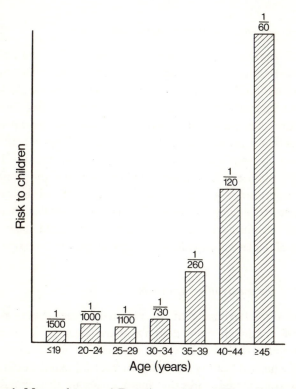

Fig. 5-6. Maternal age and Down's syndrome. (From E. Matsunaga and T. Muruyama. *Nature, Lond.* **221,** 642, fig. 2 (1969).)

that oocytes ovulated later in life are subjected to environmental influences not experienced by those ovulated early in life; for example, the process of ovulation may be impaired in the aged ovary, so that the egg tends to deteriorate before fertilization. To some extent the increase in the incidence of abnormal offspring may be peculiar to man. It has been suggested that in polytocous species (animals that have litters) intrauterine competition may eliminate abnormal conceptuses. (This is perhaps the right moment to stress that there is a need for more cytogenetic studies on developing embryos; information on these matters is given in Book 2.)

Ageing of the oocyte within the Graafian follicle

In cyclic female rats ovulation can be delayed by one or two days by treatment with appropriate drugs, such as pento-barbital sodium. Whilst the mechanism of action of the blockade has been intensively studied, scant attention was given to the possibility that the oocytes may suffer damage; recent work suggests that delaying ovulation may result in the production of abnormal zygotes and increased embryonic mortality. However, the picture is complicated by the fact that the chain of processes preceding ovulation may be disrupted, and this might contribute to embryonic mortality. In aged females there is evidence of pituitary LH deficiency, which could delay the process of ovulation, causing the oocyte to become overripe.

Ageing of the oocyte after ovulation

In general the fertilizable life of mammalian eggs is very short, probably not much more than 12 hours. Even so it tends to exceed the period within which fertilization will lead to the birth of a normal healthy offspring. In most animals the existence of an oestrous state ensures that spermatozoa are already waiting at the site of fertilization – the ampulla – before the eggs are shed (see Book 1, Chapter 3). An obvious and most important exception to this rule is the human species in which sexual intercourse may occur at any time in the menstrual cycle. Consequently if the frequency of intercourse is low, then it is more likely to occur out of phase with ovulation.

When rats and rabbits are mated or artificially inseminated late in relation to ovulation, there is appreciably more abnormal fertilization, mainly polyspermy. Normally mammalian eggs have a 'block to polyspermy' which restricts the entry of extra spermatozoa; this involves a change in either the zona pellucida, as in the rat (and in man and farm animals), or in the surface of the vitellus, as in the rabbit. The effectiveness of this blockade may be gauged from the fact that with normal time relations the incidence of polyspermic fertilization rarely exceeds 1–2 per cent, but when mating is delayed the frequency can rise ten-fold, owing, it is thought, to deterioration in the ageing eggs. Polyspermic fertilization with one extra spermatozoon leads to triploidy, which is now recognized as a common cause of miscarriage early in pregnancy. (These points are discussed further in Book 1, Chapter 5, and Book 2, Chapter 5.)

Spermatozoa

Spermatozoa may be subject to ageing both before ejaculation in the male tract and after ejaculation in the female tract, as well as under artificial conditions *in vitro*. When rabbit spermatozoa are experimentally retained in the male tract by ligation of the vasa efferentia, thereby preventing passage of testicular spermatozoa into the epididymis, fertility is maintained for 38 days and motility for as long as 60 days. Similarly, in both the rat and guinea pig motility lasts at least twice as long as the ability to fertilize. Decreasing fertility is accompanied by increase in the frequency of intrauterine mortality and fetal abnormalities. This problem has also been studied in the rabbit by testing the effect of extending the interval between semen collections. But even with intervals as long as 40 days between collections there were no significant differences in semen quality, fertilization rate or embryonic mortality that could be ascribed to collection interval. This is probably because spermatozoa are continuously eliminated from the male tract. Moreover, aged spermatozoa fare badly when in competition with fresh spermatozoa, as demonstrated in poultry and rabbits, despite the fact that they may be fertile when used alone.

Ageing of spermatozoa in the female tract may be achieved by carrying out insemination at progressively longer intervals before ovulation. The rabbit is ideal for such studies because ovulation is non-spontaneous and can easily be induced by injection of luteinizing hormone or by allowing mating with a sterile male. Rabbit spermatozoa retain their fertilizing capacity

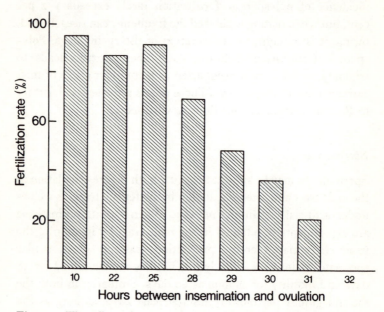

Fig. 5-7. The effect of ageing of spermatozoa on fertilization rate in the rabbit. (From J. M. Tesh. *J. Reprod. Fert.* **20**, 299, text-fig. 1 (1969).)

for 25 hours in the female tract, but thereafter a decline sets in so that after 31 hours fertility is lost (see Fig. 5-7). It appears that the main effect is at fertilization, a result that differs from what happens after ageing in the male tract.

STORAGE OF SPERMATOZOA *in vitro*

Much work has been done on the storage of spermatozoa *in vitro*, particularly of bull spermatozoa on account of the bull's economic importance in national artificial insemination schemes. The effect of length of storage and of temperature on fertility is shown in Fig. 5-8. After an unexpected initial increase attributed to the selective elimination of defective spermatozoa, fertility declines at a rate depending upon the storage tempera-

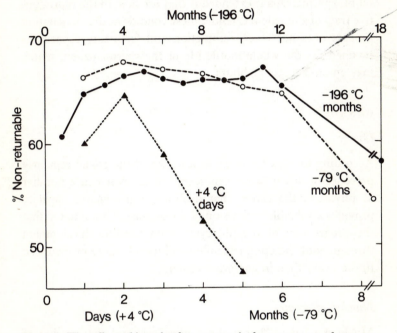

Fig. 5-8. The effect of length of storage and of temperature of storage on fertility of bull spermatozoa. (From G. W. Salisbury and R. G. Hart. *Biol. Reprod. Suppl.* **2,** 1, fig. 1 (1970).)

ture, and in the opinion of George Salisbury and Robert Hart in Urbana, this is due to increasing embryonic mortality caused by changes in the genetic information carried by the spermatozoa.

CAPACITATION

During their passage through the female tract spermatozoa of some species, if not all, undergo an as yet imperfectly understood process termed 'capacitation', which prepares them for penetration of the egg membranes (as described in Book 1, Chapter 5). Recent work suggests that once spermatozoa have become capacitated their life-span may be quite limited. In this connection it is interesting that the 'fertilizing life-span' of rabbit spermatozoa is extended if they are held in the reproductive tract of a female rat. Under these conditions the capacitation process does not reach completion and fertilizing capacity is retained for the whole motile life of the spermatozoon, which may amount to 50–55 hours.

GONADS

Ovary

According to Alex Comfort, senescence of the gonad regularly precedes or accompanies senescence of its owner in a number of phyla, to the extent that declining reproductive capacity provides a valuable expression of senescence. Thung noted that the progression of the life cycle from juvenile development through adult function to senile involution is more pronounced in the ovary than in any other organ.

Number of oocytes

Much of what we know about oogenesis and the number of oocytes present in the mammalian ovary was obtained in the 1960s, e.g. in man, monkey, guinea pig, mouse, rat, ferret, golden hamster, rabbit, cow, sheep and pig. For the majority of animals the facts are still unknown. At birth the stock of oocytes is measured in many thousands, e.g. in man estimates range from 200 000 to more than half a million, but most are destined to degenerate by a process known as atresia; the reduction in oocyte numbers from birth to 45 years is shown in Table 5-2. Only a few will actually be ovulated and of these, even in the most fecund species, no more than about 150 are likely to be fertilized and develop to term. The study of Esther Jones and

Peter Krohn in Birmingham on mice already referred to (p. 130) will serve to illustrate the relationship between age, numbers of oocytes and fertility. They found strain variations in the stage of development reached at birth, in the total numbers of oocytes present and in the rates at which the oocytes are lost (Fig. 5-9). Only in one of their strains, identified as *CBA*, did the ovary become totally depleted of oocytes long before death; at the same age, 437 days, strain-*A* ovaries contained an average of about 365 oocytes, strain-*R* III ovaries, 540, and

TABLE 5-2. Effect of age on number of oocytes remaining in human ovaries. (From E. Block. *Acta Anat.* **14,** 108 (1952).)

Age (years)	No. of cases	No. of oocytes
Birth	7	733 000
4 to 10	5	499 200
11 to 17	5	389 300
18 to 24	7	161 800
25 to 31	11	62 500
32 to 38	8	80 200
39 to 45	7	10 900

CBA × *A* hybrid ovaries about 865, suggesting the existence of hybrid vigour. Reproduction had no significant effect on the rate at which the total number of oocytes declined. The *CBA* strain lost 28 per cent of the existing oocytes every 28 days, whilst in the other strains the loss was only half this amount.

Ovarian response to pituitary hormones

In some species, such as the cow, the ovary is already capable of responding at birth to injections of follicle stimulating hormone, FSH, with marked follicular enlargement, though ovulation rarely occurs; in others, no follicular response may be evoked for weeks or even months. This variation depends upon the stage of differentiation of the ovary, which in some species is well advanced at birth. The appearance of vesicular (antral) follicles containing liquor folliculi distinguishes the onset of response. In the rabbit this stage is reached at about eleven weeks of age when the first ovulations can be induced – still well before the animal is capable of supporting a pregnancy. Ovarian response rises to a maximum at 5–6 months in the rabbit and

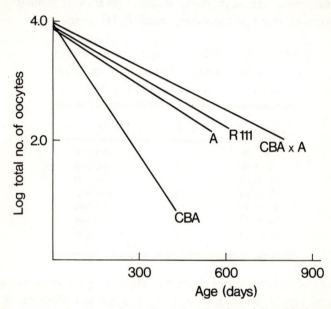

Fig. 5-9. Combined regression lines showing the relationship between age and total numbers of oocytes in various strains of mice. (From E. C. Jones and P. L. Krohn. *J. Endocr.* **21**, 469, text-fig. 9 (1961).)

then levels off before declining significantly with age, as illustrated in Fig. 5-10. Likewise the natural ovulation rate declines by 30 to 40 per cent in aged does, presumably reflecting the decline in the number of large follicles. A similar decline also occurs in the mouse (Fig. 5-11). Eggs obtained from immature animals are capable of being fertilized and, if transferred to a favourable environment, of developing into normal, healthy offspring.

Precocious puberty in girls is often associated with a lesion of the hypothalamo-pituitary region, which presumably leads to early release of pituitary hormones; evidently in man too the ovary is capable of responding to hormone stimulation well before the normal time.

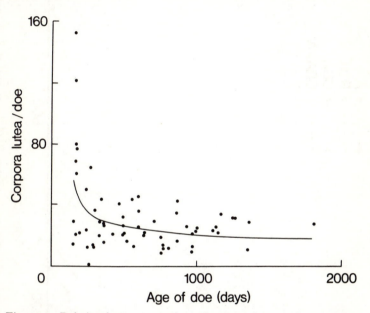

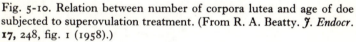

Fig. 5-10. Relation between number of corpora lutea and age of doe subjected to superovulation treatment. (From R. A. Beatty. *J. Endocr.* **17,** 248, fig. 1 (1958).)

Endocrine function

Apart from its gametogenic role, the ovary is an important endocrine organ, being responsible for the secretion of oestrogen and progesterone, but this function waits upon the ripening of follicles, ovulation, and the development of corpora lutea. During the period of lowered fertility late in the life-span there is no direct evidence that the secretion of either of these hormones is significantly reduced. In fact, reproduction seems to end well before endocrine function becomes impaired. The

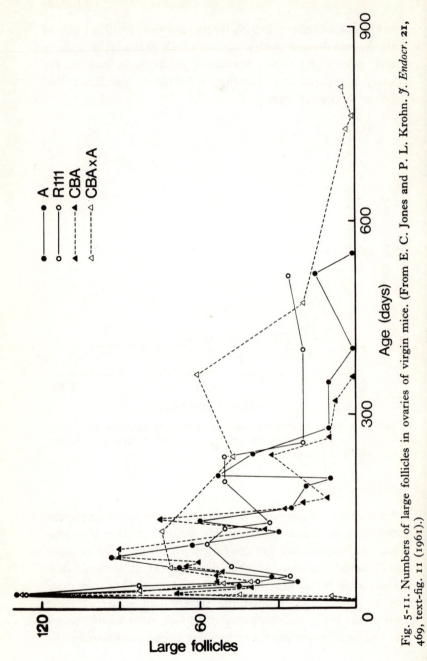

Fig. 5-11. Numbers of large follicles in ovaries of virgin mice. (From E. C. Jones and P. L. Krohn. *J. Endocr.* **21**, 469, text-fig. 11 (1961).)

latest ages at which young corpora lutea and large follicles were found in virgin mice are given in Table 5-3.

Thung, in reviewing 'Ageing changes in the ovary', presents as evidence for the decreasing efficiency of ovulation and corpus luteum formation in ageing mice the frequency with

TABLE 5-3. Latest ages at which young corpora lutea and large follicles were found in virgin mice (from E. C. Jones and P. L. Krohn *J. Endocr.* **21,** 469 (1961).)

Strain of mouse	Age at which young corpora lutea present (days)	Large follicles present (days)
CBA	320	360
A	525	561
R III	100	589
CBA × A	480	852

TABLE 5-4. The number of corpora lutea and the percentage of luteinized atretic follicles (corpora lutea enclosing an oocyte) in serially sectioned mouse ovaries at different ages (From P. J. Thung. In *Structural Aspects of Ageing*, Ch. 9. Ed. G. H. Bourne and E. M. H. Wilson. London, Pitman (1961).)

Age in months	No. of ovaries sectioned	Av. total no. of corpora lutea per ovary	% of luteinized atretic follicles
	Strain C 57 BL		
16–17	12	7.9	5
18–21	12	5.6	28
	Hybrids $F_1(O_{20} \times DBA_f)$		
14–19	17	14.7	2
20–25	32	4.8	36
26–30	28	1.0	32

which follicles undergo luteinization without previous ovulation (see Table 5-4). Studies on ageing mice suggest that the amount of circulating LH becomes inadequate or ceases altogether, and it may be that this is attributable to ageing of the hypothalamus.

THE GENITAL TRACT

In ageing female mammals declining ovarian function is shown by loss of weight of the organ and a reduction in numbers of both follicles and corpora lutea (see Table 5-5). An extreme condition is found in women at the menopause when ovarian function fails, and subsequently the genital tract atrophies.

Ageing may also lead to changes of a pathological nature, some of which are gross and immediately recognizable, whilst others are at the cellular level and require microscopic study and expert interpretation. In the first category a good example is the condition known as 'cystic ovary', the incidence of which is clearly dependent on age in rodents, and possibly in other orders, too, though comparative data are lacking. Information on the incidence of the condition in rats and hamsters, is given in Table 5-5, and in the Mongolian gerbil (*Meriones unguiculatus*) in Fig. 5-12. Among females aged 200 to 400 days, only 5 per cent had cystic ovaries, whereas in those examined at 600 to 900 days 73 per cent were found to have cysts with an increasing proportion having both ovaries affected. Sometimes the cysts assumed massive proportions, accounting for up to one-sixth of the animal's weight. The ovarian tissue becomes stretched to a thin film on the surface of the cyst and conception fails. However, when only one ovary is affected the other may continue to function normally, showing that the effect is quite local. An example of cystic ovary in the gerbil is shown in Fig. 5-13.

The second kind of pathological change seen with ageing in the ovary includes certain forms of cellular degeneration (ceroid and amyloid), the accumulation of so-called 'wheel' cells and recognizably sterile follicles, and the formation of testis-like tubules and ingrowths of the germinal epithelium. Ovaries from

TABLE 5-5. Comparison of ovarian changes between adult and aged rats and hamsters (From A. P. Labhsetwar. *J. Reprod. Fert.* Suppl. 12, 99 (1970).)

	Rat		Hamster	
	Adult (3 months)	Aged (22–32 months)	Adult (3 months)	Aged (21 months)
No. of animals/group	8	28	5	5
Ovarian wt (mg/ovary/animal)	49±2*	36±4	Not reported	Not reported
No. of follicles/ovary	29±3.9	9±0.9	42.8	12.4
Mean no. of c.l.†/ovary	15±1.4	8±1	4.8	3.0
% animals with cystic follicles	0	43	0	20
% animals with 'wheel' cells in ovaries	0	75	Not reported	Not reported

* mean ± S.D.
† corpora lutea.

ageing multiparous mice were found to have fewer histological abnormalities than those from ageing virgin mice. Ovarian cancers are relatively uncommon; in ageing rats and mice the incidence of primary ovarian tumours is only about 0.03 per cent and 0.2 per cent, respectively.

In women, the genital tract often becomes diseased to the

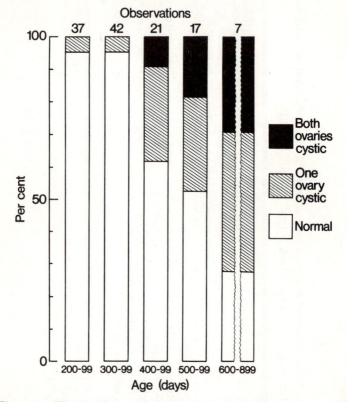

Fig. 5-12. The incidence of cystic ovaries according to age in the Mongolian gerbil. (The last column is split because it covers a much longer time interval than the others.)

extent that major surgery is required. A recent Dutch survey based on more than 4000 women in the 42–62 years age group revealed that one-sixth had undergone surgery of the uterus or ovaries.

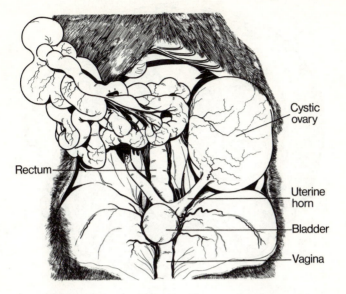

Fig. 5-13. 'Cystic ovary' in the Mongolian gerbil.

In the rabbit, too, the uterus is particularly prone to disease, and tumour formation is common. There are two reports of 80–90 per cent of does having diseased uteri by 5 to 6 years of age compared with less than 5 per cent at 2 to 3 years (see Table 5–6). Before tumours form, the endometrium exhibits a

TABLE 5-6. Incidence of diseased uteri relative to age in the rabbit (From C. E. Adams, *J. Reprod. Fert.* Suppl. **12,** 1 (1970).)

Doe's age (years)	No. does	Condition of uterus					
		'Normal'		One horn diseased		Both horns diseased	
		No.	%	No.	%	No.	%
4 to 4.9	22	10	45	7	32	5	23
5 to 6	29	2	7	1	3	26	90

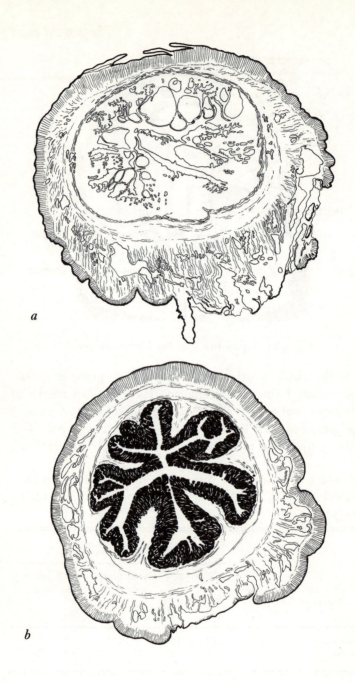

a

b

cystic appearance (cystic glandular hyperplasia) and fails to respond to progesterone, as shown in Fig. 5-14. The condition causes sterility, for embryos cannot survive in such an environment. Since cystic hyperplasia also occurs commonly in women before the menopause, it may well be responsible for the declining fertility that characterizes this phase, but we have no definite information on the effects of uterine ageing in monotocous species, including man.

According to Marcus Bishop the degenerative changes that occur in the testis 'appear to be related to no other obvious factor than advancing age itself'. Most commonly we see a reduction in the size and activity of the seminiferous tubules.

FOLLICULAR CONSTANCY

When one ovary is removed in polytocous animals, the remaining one compensates by shedding roughly the same number of eggs as the two had done before, and fertility, at least initially, does not suffer. This fact established by John Hunter nearly two centuries ago has been found true for all those species so far examined, including the opossum, golden hamster, rat, mouse, rabbit and Mongolian gerbil. Moreover, if all but a fragment of ovarian tissue is removed, that fragment can hypertrophy to approximate the size of two normal ovaries. These observations led Lipschütz to formulate his 'law of follicular constancy', which he stated as follows: 'The number of primary follicles which enter into follicular development, the degree of follicular ripening which is attained, and the further fate of the follicle depend not upon the total number of primary

---

Fig. 5-14. *a* Section of uterus, 60 h *post coitum*, from a multiparous rabbit doe aged 50 months. Note extensive cystic enlargement (hyperplasia) and lack of endometrial growth. (From C. E. Adams. In *Preimplantation Stages of Pregnancy*, p. 345, fig. 11. Ed. G. E. W. Wolstenholme and M. O'Connor. Churchill, London (1965).)
b Section of uterus, 96 h after first mating, from a doe aged 7 months. Note extensive endometrial growth. (Same source as *a*, fig. 10.)

follicles present, but upon general internal factors outside the ovary.' Basically, the mechanism of follicular constancy depends on the fact that the amount of FSH released from the pituitary is increased initially (owing to absence of 'negative feedback' – see Book 2, Chapters 1 and 2), and then declines to the normal level when an appropriate amount of active ovarian tissue has been restored. The compensatory hypertrophy fails in aged rats whose ovaries can still respond to gonadotrophin, because the FSH secretion rate declines.

Information as to whether hemi-ovariectomy in women hastens the onset of the menopause is scarce, though it is known that in mice the rate of loss of oocytes in a single ovary is substantially greater than would be expected to occur in one ovary of a normal animal. Removal of one ovary in the mouse does disturb the normal pattern of the oestrous cycle, but the cycles cease earlier in life than in control animals (see Fig. 5-15).

Experiments on mice designed to test the long-term effects of semi-spaying on reproductive performance have produced some unexpected and potentially far-reaching results. Although at first the performance of the one-ovary mice equalled that of their controls, after about the sixth pregnancy they ceased to reproduce. In the mouse, unlike the pig, migration of eggs from one uterine horn to the other is a very rare event and pregnancy is normally confined to one horn which, therefore, carries double its normal load. The uterus appears to become prematurely exhausted and although the exact point of breakdown has not been defined, impairment of the vascular supply is suspected. We may be witnessing the acceleration of a normal process of ageing.

When the experiment was repeated on rats a similar overall result was obtained though the pattern of reproduction was slightly different. In this case, too, the sixth pregnancy represented a dividing line, but now reproduction fell to a low level rather than ceased altogether. In fact, the mean age of the semi-spayed rats when they produced their final litter equalled that of their controls. However, the one-ovary rats did produce

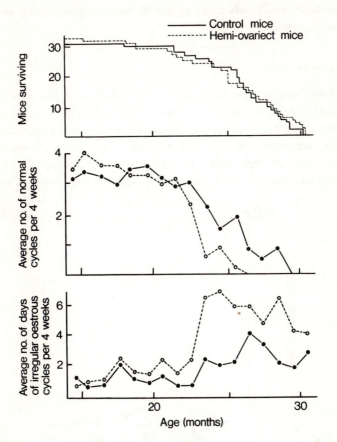

Fig. 5-15. Ageing changes as exemplified in the oestrous cycles of normal and unilaterally ovariectomized F_1 ($O_{20} \times DBA_f$) mice. The animals were allowed to live their maximum life-span. The average number per animal of normal and irregular cycles was calculated over 4-week intervals. (From P. J. Thung. In *Structural Aspects of Ageing*, fig. 10. Ed. G. H. Bourne. Pitman Medical, London. (1961).)

significantly fewer litters, reflecting a longer interval between successful pregnancies. This feature is also reminiscent of the effect of ageing on reproduction. When representatives of the two groups were examined in their sixth pregnancy, the number of ovulations was found to be similar, but the level of embryonic mortality was nearly twice as high in the one-ovary group as in the controls.

CLIMACTERIC

Except possibly for the mouse, more is known of the terminal years of reproductive life in man than in any other mammal. A distinguishing feature, shared possibly by few other species, is the menopause, which proclaims the final decay of ovarian function. At this point menstruation fails completely – the outward sign of the menopause – although the uterus is still

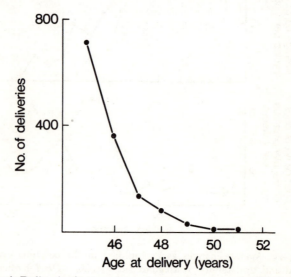

Fig. 5-16. Deliveries in women aged more than 44 years, England and Wales, 1967. (From W. J. A. Francis. *J. Reprod. Fert. Suppl.* **12**, 89, text-fig. 3 (1970).)

fully capable of responding to hormone therapy. Therefore, in menopausal women who are taking oral contraceptives, menstruation continues. Clinically, the menopause is especially significant because it often involves endocrine, somatic and psychic changes which together constitute the climacteric, or change of life. Clinicians suggest that a climacteric may also occur in some men, though when it does it takes place later in life than in women, between the ages of 55 and 65 years.

There is a well-defined trend for the menopause to occur later in life, in spite of the fact that menarche shows the opposite tendency. In Britain the average age at the menopause is now 50 years which is four years later than it was a century ago. Many women menstruate regularly till the age of 54 or 55 years, but conception after the age of 50 is quite rare. Thus, the post-reproductive period of life now spans two decades or more. In England and Wales in 1967, the incidence of pregnancy in women more than 44 years of age was 0.1 per cent, as shown in Fig. 5-16. Even in the Hutterites, a strictly religious sect living in Alberta and South Dakota, who do not appear to practise contraception, childbearing ceases at about 49 years of age. Undoubtedly, the incidence of pregnancy in ageing women is now limited by socio-economic trends and the use of contraceptives. Data from the United States show that, in 1910, 11 per cent of women between 45 and 49 years of age had children under 5 years old, whereas in 1950 this figure had shrunk to 5 per cent.

Ageing affects many functions and nowhere is this more evident than in reproduction and the processes of development. It is not only an inevitable change but also a continuous one, going on throughout life; as a popular aphorism puts it, we start to age as soon as we are born. Even this is not the whole truth, for with the occurrence of preovulatory ageing, it would seem that we may begin to age before we begin!

SUGGESTED FURTHER READING

The Biology of Senescence. A. Comfort. London; Routledge and Kegan Paul (1956).
The causes of ageing. A. Comfort. *Science Journal* **1,** 70 (1965).
The relationships between age, number of oocytes and fertility in virgin and multiparous mice. E. C. Jones and P. L. Krohn. *Journal of Endocrinology* **21,** 469 (1961).
The reproductive life-span. P. L. Krohn. *Proceedings of the Fifth International Congress of Animal Reproduction, Trento* **2,** 33 (1964).
Ageing and reproduction. Ed. J. S. Perry. *Journal of Reproduction and Fertility* Suppl. **12,** 1. (1970).

Ageing in reproduction

Gamete ageing and its consequences. G. W. Salisbury and R. G. Hart. *Biology of Reproduction* Suppl. **2**, 1 (1970).

Effect of maternal age on reproductive capacity. G. B. Talbert. *American Journal of Obstetrics and Gynecology* **102**, 451 (1968).

The Biology of Ageing. Ed. W. B. Yapp and G. H. Bourne. Symposium of the Institute of Biology, 6. London (1968).

Ciba Foundation Colloquia on Ageing, vol. 1. Ed. G. W. Wolstenholme and M. O'Connor. London; Churchill (1955).

Ageing changes in the ovary. P. J. Thung. In *Structural Aspects of Ageing.* Ed. G. H. Bourne and E. M. H. Wilson. London; Pitman Medical (1961).

Index

ABO blood-group incompatibility, 121
accessory organs, female, ageing of, 130
acrosome, as source of sperm antigens, 103
Addison's disease, and autoantibodies against ovary, etc., 110
ageing, and capacitation, 140
and endocrine function, 143–6
and reproduction, 128–56
decline in fertility with, 128
genital tract, 146–51
agglutinin, as type of antibody, 95
altricial young, 54
anaphylaxis, uterine, during intercourse, 111
antibody, 94–100
ABO blood group, 96
against hormones, 125, 126
against spermatozoa, 111, 112
anti-Rh, 120–4
as experimental tool, 124–6
chemical nature and properties of, 95–100
circulating, 94, 103–7, 110–12
five classes of, 96–8
'fragments' of, 96, 97, 117, 119
in females, against spermatozoa and seminal plasma, 110–12
in pregnancy, 112–16
uterine, 111, 113
antigen, ABO, 108
blood group, 107, 108
characteristics of, 94, 95
rhesus, during pregnancy, 120–4
'sperm-coating', 109
transplantation, 107, 108, 112, 113, 116
antitoxin, as type of antibody, 95
atresia, of follicles, 140
autoimmune aspermatogenesis, 101–5
autoimmunity, to ovaries and adrenals, 110
to spermatozoa in men, 105–7

to spermatozoa in women, 110–12

basement membrane, of testis tubule, as barrier to antibodies, 104
behaviour,
maternal, 54–68
eating placenta, 56, 59
eating young, 56
influence of hormones on, 61–3
licking young, 56, 58
mother–young interaction, 58–61, 65–8
nest-building, 56, 58
retrieving young, 57, 58
role of experience, 64
role of young, 63–4
towards young in society, 64–8
sexual, 35–54
attractiveness, 48, 49
control by CNS, 44–6
courtship, 35–6
differentiation of, in primates, 50–2
during menstrual cycle, 48
effect of castration on, 39–42, 43, 44, 47, 49, 50
effect of hormones on, 39–42, 43–5, 49, 50
feminine, 42–4
heterotypical, 42–3
homosexual, 52–4
homotypical, 42–3
in primates, 47–54
masculine, 42–4
mating, 35–6, 46
oestrous, 36–9, 47, 49
receptivity, 47, 49
role of experience, 46–7
smell, 37, 39, 48
visual cues, 36–8, 49
vocalization, 39
territorial, rabbits, 88
Bruce effect, 87

capacitation of spermatozoa, ageing and, 139, 140

Index

carrying capacity, of habitat, 13, 16
cellular immunity, 94, 103
climacteric, 154, 155
cold, effect on reproduction, 76
colostrum, and transfer of maternal antibodies, 116–19
communication, between mother and young, 58–61
complement, and antibody action, 96
Cushing's syndrome, and autoantibodies against ovary, etc., 110
'cyclic' mammals (in breeding intensity), 91–3

density of animals, effects on reproduction, 89–91
diet, effect on reproduction, 78–84
Down's syndrome (mongolism), 134, 135

egg, abnormal in old animal, 130–2
ageing, 131, 133
failure of cleavage with age, 132
electrophoresis, for separation of antigens, 100
elephant, 12–21
birth rate, 16
body temperature, 20
calving interval, 13–15
clitoris, 20
corpora albicantia, 18
corpora lutea, 17, 18
cropping, 15–17
density-dependent mechanisms, 13, 15
endometrium, 17, 18
epididymis, 20
gestation length, 13, 17
lactational anoestrus, 18
life span, 14
male accessory organs, 20
mating, 18, 19
mortality rate, 13–15
musth, 21
oestrous cycle, 17, 18
ovulation, 17, 18
parturition, 18
penis, 19
placental scar, 18
population density, 14
population explosion, 13, 16
pregnancy, 18

progesterone, 17
puberty, 13–15, 19
rate of growth, 13, 15, 19
seminiferous tubule diameter, 19
survivorship curve, 14–16
temporal gland, 20, 21
testes, intra-abdominal, 13, 19, 20
testosterone, 21
vagina, 19
volume of ejaculate, 20
weight, 19
embryo, mortality, with ageing of spermatozoa, 131
energy requirements, for reproduction, 83–5
enhancement facilitation, 114, 115
environmental effects on reproduction, 69–93
equine hybrids, 21–32
hinny, 29
horsebra, 29
infertility, 29, 31
mule, 21, 24, 29, 30
zebronkey, 29–31
zebrorse, 29, 30
erythroblastosis, 120–4

fertilization, abnormal, with ageing oocyte, 136, 137
failure, with ageing, 132
fetus, as a foreign graft, 112–16
as source of foreign antigens for female, 109
'low-dose tolerance' of, in mother, 114, 115
fibrinoid, as protection for fetus, 115
flushing, of ewes to produce twins, 80
follicular constancy, and ageing, 151–5
gamete, ageing of, 133–8
genital tract, changes with age, 146–51
gerontology, 128, 129
globulin, as nature of antibody, 95
transmission through cells, 122, 123
gonad, ageing and, 140–6

haemolytic disease of newborn, 120–4
habitat, carrying capacity, 13, 16
hibernation, reproduction and, 76, 77

Index